The Other Price of Britain's Oil

A Volume in the
Crime, Law, and Deviance
Series

The Other Price
of Britain's Oil

Safety and Control in the North Sea

W. G. CARSON

Rutgers University Press
New Brunswick, New Jersey

First published in the USA by
Rutgers University Press, 1982

Library of Congress Catalog Card Number 81-85459
ISBN 0-8135-0957-2

First published in Great Britain by
Martin Robertson and Co. Ltd., 1982

Printed in Great Britain

For Ruth, Christopher and Siobhan

Contents

Preface

This study grew out of a fortuitous combination of circumstances – my long-standing interest in the field of British factory legislation and the fact that, since 1974, I have been living and working in Scotland, off whose shores substantial quantities of oil had already been discovered by the time I arrived. At that particular time, there was much talk not only about the unexpected windfall which had now come Britain's (quite a few felt it should be Scotland's) way, but also about the possibility that, as with other sudden discoveries of unanticipated wealth, the 'oil rush' might turn out to be a rough and tough affair in which questions of safety were pushed aside in the pursuit of riches previously undreamed of. Sensing that here there might be something of more than passing interest to someone interested in the field of safety legislation, I decided to investigate. It would be a lie to pretend that I came to the subject through any prescience which told me that this particular topic was one which encapsulated crucial sociological issues or matters of vital theoretical concern to the sociology of law. All I wanted to do was to find out how the law relating to safety was operating in the North Sea.

During the latter part of 1976, the Oil Panel of the Social Science Research Council provided me with a small grant to establish whether research in this particular field was feasible. Having presented a brief report suggesting that such was indeed the case, I approached the Scottish Home and Health Department for assistance, and from the beginning of 1978 a further grant was made available. I would like to record my gratitude to both these bodies for the support which they have given in making this project possible.

In the course of the research, horizons broadened, ideas changed and, as readers will see, I came to appreciate the need for an analysis which was couched in terms much wider than those of the nitty-gritty of offshore safety legislation or, indeed, of North Sea oil itself.

Research methods also had to be tailored to meet the requirements of a somewhat unusual case, in which the geographical location was remote, co-operation not always readily forthcoming and the subject matter of a more than slightly sensitive nature. Systematic and structured interviewing proved to be of only limited value; access to some crucial kinds of official data turned out to be extraordinarily difficult; and, of course, what used to be applauded as a respectable alternative, participant observation, was possible only on a restricted scale. As the project progressed, I came to define certain aspects of it as an exercise in investigative sociology, and, as is commonplace with work of that kind, I have at many points abided by the protocols pertaining to the confidentiality of my sources. Throughout, however, I have attempted to avoid embellishment of the accounts and information offered to me.

Inevitably, a project of this kind leaves me deeply indebted to many people. In particular, the ready assistance of the Scottish Crown Office and of the Procurator Fiscal's office in Aberdeen has been invaluable, not least because it allowed me an entrée to detailed information about offshore accidents which might not otherwise have been available. Jacqueline Tombs of the Scottish Home and Health Department has been consistently helpful and supportive throughout, and I am also grateful to those officials of the Department of Energy, the Health and Safety Executive and the Department of Trade who spared the time from other pressing matters to talk to me. Trades union officials also gave freely of their time, as indeed did safety officers and other executives in the oil industry. Offshore workers who tolerated the tedium of lengthy tape-recorded interviews provided valuable insights into the world of offshore safety as perceived from the sharp end, while numerous others were only too willing to share their experiences, as well as their 'carry-outs', in the course of protracted train journeys between Edinburgh and Aberdeen. Workers and management even tolerated my presence on several installations with great courtesy and good humour.

On the academic side, I am extremely grateful to my colleagues in the Department of Criminology at the University of Edinburgh for bearing with me in my possibly peripheral criminological preoccupation. At different times and in different ways Derick McClintock, Peter Young, David Garland and David Jenkins have all provided me with support when it was badly needed. In the early stages of the project, Hilary Idzikowska contributed invaluable assistance, particularly with the time-consuming business of interviewing and the collection of accident data. Other colleagues, in Edinburgh and elsewhere, have also been unstinting in their assistance. Professor J. K. Mason of the Edinburgh Department of Forensic Medicine kindly checked part of my manuscript for medical *faux pas*, while Professor A. W. Hendry scrutinized chapter 3 for any obvious gaffes in the field of civil engineering. Professor R. Black of the Department of Scots Law read the entire manuscript with an eye to saving me from doing violence to the law of Scotland, as well as from some other uninvited consequences of working in the area of offshore safety. Pat Carlen and Ian Taylor read several chapters in order to reassure me when I became sociologically fainthearted. David Nelken of Edinburgh and Jason Ditton of Glasgow have read the entire book most assiduously, and I can best thank them both by saying that it is on them that I have leaned most heavily during the time this study has been in preparation. Jeanette McNeill typed the manuscript with almost unerring accuracy, with no mean academic eye of her own and with a degree of enthusiasm which did not readily countenance delay or second thoughts. When assistance with this respect of producing the final manuscript was required, Helen Dignan and Lorna Paterson willingly came to my rescue.

Finally, I have always thought that some acknowledgements of indebtedness at the front of books are placed there as a matter of formality. In the present instance, however, the encouragement and support provided by Michael Hay of Martin Robertson & Co. have gone so far beyond the call of publishing duty that prefatory mention can scarcely capture the extent of my indebtedness. At home, the catalogue of lost weekends, disturbed nights, preoccupied days, unbuilt train sets and missed fishing trips which my family has endured in the course of this book's preparation now tells me that thanking families is no formality either. With an apology for ever having thought it otherwise, I would therefore like to thank Ruth, Christopher and Siobhan for putting up with me while I have been doing my particular thing.

Kit Carson
Duddingston Village,
Edinburgh, 1981

1

Introduction

As one listened, one had a kind of feeling of floating almost
imperceptibly into another new and exciting phase of this brave
new world. . . . And here we are, talking about the tremendous
wealth under the seas. Imagination simply boggles. . . .
Sir Frank Soskice, MP,
Second Reading Debate, Continental Shelf Bill, 1964

According to Eric Hobsbawm's classic history of industry and empire,
Britain was a much more comfortable and entertaining place to live
during the early 1960s than it had ever been before.[1] Prophets of doom
notwithstanding, he maintained, Britain was not a 'paralysed or sinking wreck', but a country of great potentialities and resources, both
human and technical, if only they could be effectively harnessed. And
yet, somewhat to his apparent relief, people were not content, even if
they were immeasurably better off. Gaps between aspirations and
reality remained to generate unease, and national self-esteem had
reached a low ebb:

> Man does not live by gas central heating alone, even though the
> assumption of the advertisers, the most effective mass ideologists
> since the decline of the churches, seemed to be that he should. Hope
> and pride had grown dim. 'Few', writes A. J. P. Taylor, 'now sang
> Land of Hope and Glory. Few even sang England Arise.' And yet, if
> there was not much scope left for the first of these songs, there was
> still plenty left for the second.[2]

One ray of hope was, however, beginning to be just dimly perceptible
from British shores at this time. Following the discovery of massive
deposits of natural gas in Holland at the end of the previous decade, a
number of companies, including most of the multinational oil corporations, had been carrying out magnetic and seismic surveys in the North
Sea, and by the end of 1963 the results were such as to justify embarking upon offshore drilling to establish whether early hopes would be

realized in practice. With the provision of the requisite legal framework by the Continental Shelf Act 1964,[3] exploration soon got under way off the eastern coast of England, efforts being rewarded in the autumn of the following year when British Petroleum (BP) discovered major gas deposits in what became known as the West Sole field. Further gas finds in the same general area followed rapidly, with the Leman Bank, Indefatigable, Hewett, Viking and Rough fields all being discovered by mid-1968.[4] Production started in 1967, and in the same year a White Paper on fuel policy predicted with some confidence that 1972 or 1973 would see it running at a rate of around 3000 million cubic feet per day,* nearly three times the UK's then current consumption of town gas.[5]

To a country whose people, inasmuch as they had thought about the matter at all, had tended to identify natural gas with the bountiful life to be enjoyed in what were then far-off lands like Canada, these developments must have come as a highly unexpected windfall. But still better was to come, even if in a form likely to stimulate patriotic renderings of 'Flower of Scotland' rather than 'England Arise'. Exploration having moved northwards into the stormier, deeper and much more hostile waters off the coast of Scotland, oil was discovered by Amoco in September 1969. Whereas gas had stirred some glimmerings of optimism in the 1960s, oil was now to provoke nothing far short of euphoria in the 1970s. Fields like Forties, Brent and Ninian were to become almost household names, while Hamilton Brothers scored an historic triumph by bringing the first of Britain's oil ashore from the Argyll field in June 1975.[6] Just over a year earlier, the Secretary of State for Energy had been able to tell Parliament that North Sea gas was already meeting around 90 per cent of the country's gas consumption, which had tripled since 1965, while there was now 'a very good chance' that net self-sufficiency in oil would be reached by 1980.[7] And all this at a time when Britain had not only become dependent on oil for almost 50 per cent of her total primary fuel consumption but was also reeling under the imapct of price rises imposed by the Organization of Petroleum Exporting Countries (OPEC) at the end of 1973.

In the event, the optimism of 1974 was to prove fully justified, at least in terms of offshore progress. Net self-sufficiency in oil was

---

* The units of measurement used in many of the sources from which information has been gathered during the course of research for this work are Imperial. Apart from references to the 500-metre safety zone around North Sea installations that is enshrined in law, it has been considered expedient to retain Imperial units throughout the book in the interests of accuracy and consistency. However, a brief list of approximate SI equivalents is provided on p. 310 for the benefit of readers to whom these may be more familiar.

reached during the summer of 1980, 79.2 million tons being produced during that year. At 31.7 million tons of oil equivalent, natural gas was meeting around 78 per cent of the country's primary gas demand, while offshore oil and gas were supplying 40 per cent and 16 per cent of the UK's total primary fuel consumption respectively.[8] Total possible reserves of oil on the Continental Shelf were being estimated at between 2141 and 4281 million tons, of which 259 million had already been used up.[9] By 1980 as well, the Department of Energy's annual report on the development of the oil and gas resources of the United Kingdom (*The Brown Book*) had even been able to give up its earlier and persistent preoccupation with the advantages accruing to Britain's balance of payments from the Continental Shelf. Instead, the emphasis was now upon revenues, which, during the financial year 1980–1, were provisionally estimated at some £3.8 billion, a figure approximating nearly one-third of the yield from value added tax and more than one-sixth of the entire revenue collected by government in the form of income tax.[10] A Hobsbawm taking stock of Britain again in the early 1980s would probably have to concede that, despite the fairly unremitting gloom of the twenty years that had elapsed since the exercise was last undertaken, one great British resource had been effectively harnessed – the wealth of oil and gas beneath the North Sea. Nor is the ordinary Britisher allowed to forget it. While the volume may be turned down on the politicians' almost ritualistic allusions to the subject, the advertisers now nightly celebrate on the nation's television screens the great job done for the United Kingdom by the multinational oil corporations operating on the Continental Shelf.

Behind the almost astonishing success story of the North Sea there lies a tale of remarkable ingenuity, technological sophistication and engineering temerity. Offshore fields, particularly the oilfields off the Scottish coast, may be several hundred miles from land and may lie beneath waters more than 600 feet in depth, where winds can be ferocious and the once-in-a-hundred-years wave of well over 100 feet has nonetheless to be expected daily. While drilling for oil in such a setting may itself be an awesome undertaking, it pales into relative insignificance alongside what follows the discovery of a viable field. The platforms which have to be built, towed out, installed and then completed *in situ* are vast, inviting comparison in terms of the colossal and description in terms of the superlative. Thus, for example, Chevron is reported to have described the central platform for its Ninian field as 'the biggest thing that has ever been moved on earth', while more recently a television documentary conveyed some sense of the task involved in towing out another platform by equating its size with that of the British Houses of Parliament.[11] Throughout the entire operation, from initial drilling to final production, moreover, an enormous logistical exercise is involved in supplying installations by ship and by helicoptering a workforce of over 10,000 to and from the shore at the beginning and end of work shifts, which usually last for two weeks. Dyce Airport in Aberdeen is often described as Europe's busiest heliport. If the transport of personnel and material presents logistical problems, the conveyance of the oil and gas themselves offers a formidable challenge to the industry's technological and engineering expertise. Thus far, nearly 1600 miles of pipeline has had to be laid beneath the North Sea,[12] while terminals to receive tankers and pipelines have had to be constructed in remote locations such as Sullom Voe in Shetland and Flotta in Orkney.

Needless to say, this has all cost quite a lot of money. By the end of 1980, it is calculated, total investment will have reached about £21 billion in 1980 prices, with expenditure of a further £4 billion on exploration.[13] During the second half of the 1970s, annual offshore-related investment was reckoned to be running at around 20–25 per cent of total UK industrial investment,[14] a degree of concentration possibly unrivalled since the great railway-boom over a hundred years earlier. In late 1979, it was estimated that North Sea investment costs were averaging US $8000 per daily barrel produced, and that this figure might rise to $14,000 by the end of the century as companies are forced to fall back on the smaller, more costly reservoirs which will remain.[15] According to Guy Arnold, even in 1975 a semi-submersible drilling rig could cost £25 million to build and £30,000 daily to operate, while Shell have calculated the cost of their North Sea exploration and production activities at £13 per second. Although Arnold's conclusion that 'the North Sea represents the most costly oil and gas producing area in the world' did not take account of the still greater projected costs of extracting oil from the Arctic or from tar sands, there is no doubt that exploitation of the resources lying beneath the UK Continental Shelf has been a very expensive business indeed.[16]

To some observers, these are not the only costs which have been high. As we shall see at a later point, for example, it is by no means agreed that the price paid by the British economy as a whole has been a reasonable one, not least because of oil's impact on the value of the pound. Equally, it is arguable that the dramatic turn-about in Britain's oil fortunes has been accomplished only at the price of an inordinately high level of penetration by overseas, and particularly US capital. Nor is it established to the satisfaction of all concerned that the British Exchequer has not to a considerable extent been fleeced in the process.[17] On a narrower front, it is not universally accepted in Scotland that the oil industry's creation of short-term employment opportunities, often with considerable disruption to an established, if not so prosperous, way of life, will ultimately turn out to be a benefit rather than a penalty.[18] Least of all, perhaps, is there consensual endorsement of the view that the risk of a major environmental catastrophe associated with North Sea oil operations is a price worth paying.

This book is about another price which has been paid for Britain's oil, the price in terms of death and injury among the workers who have been employed offshore. Indeed, the starting point for my thesis about the regulation of offshore safety is that this price has been inordinately high. To be sure, the British sector of the North Sea has thus far experienced no major catastrophe on a scale to compare with the Alexander Kielland disaster, which cost the lives of 123 workers in the Norwegian sector in March 1980. Nor has there, thus far, even been a repetition of the UK sector's one serious multi-death accident, which took place in 1965, when the Sea Gem, a jack-up drilling rig, collapsed with the loss of thirteen lives. Throughout the years which have elapsed since exploration for gas and oil first began, however, a relentless (though now, thankfully, declining) price in death and injury has nevertheless been exacted in the race to get Britain's offshore wealth ashore. To date, well over a hundred North Sea workers have been killed and at least four times that number seriously injured. Moreover, while these totals may not seem excessive in absolute terms, they compare very badly, I will argue, when set alongside the record for other occupations acknowledged to be highly dangerous.

Chapter 2 will examine the statistical basis for this argument in some detail, even if the inadequacies of North Sea casualty figures counsel a degree of caution about the firmness of the conclusions to be drawn. No such caveat need be entered, however, in relation to answering another obvious question thrown up by the offshore safety record – why do such accidents happen? Here we encounter one of the most abiding of North Sea myths, namely, that the price which has been paid in casualties is simply the by-product of operations which take place at the very frontiers of technology and in adverse climatic

---

* Shortly before this book went to press, two helicopters engaged in North Sea operations crashed on successive days in August 1981, killing one and thirteen people respectively. The causes of these accidents are as yet unknown.

conditions. Consoled by such a belief, the catalogue of offshore injuries can all too easily be accepted as inevitable and, if the economic priorities of North Sea oil are accepted, as necessary. Using data from fatal accident files and other sources, chapter 3 will challenge this dominant image of offshore danger and will suggest that the vast majority of accidents result from relatively conventional causes, which are comprehensible to anyone acquainted with the mundane world of industrial safety onshore. No less important, the evidence adduced in that chapter indicates fairly clearly that, contrary to the view which would see the technological complexity and operational inevitability of offshore hazards as leaving only a peripheral role to law in the maintenance of safe working conditions, there is no reason why legal regulations should not, in principle, play just as crucial a part in this context as they do onshore. To be sure, this does not mean that I am about to embark upon an argument which suggests that law could solve all the problems of offshore safety. There is too much empirical evidence from other work settings and too much theoretical doubt about the potentialities of law itself to countenance such an idealistic view. But the best can too easily become the enemy of the good, and I do believe that law could have accomplished more good than it did in the North Sea, and that its limited success in this respect is not traceable to the unique exigencies of offshore operations as such.

Against this background, much of the remainder of the work will be devoted to an examination of the legislative and enforcement response to the issues of safety raised by the activities taking place on the Continental Shelf. In the first of these respects, it will be suggested that the reaction of the legislature was laggardly, highly deferential to the industry and constantly at pains to avoid fettering the operations which alone could establish the extent of Britain's new-found wealth and turn it into a realizable asset. Furthermore, I will argue, these tendencies carried over into the administrative arrangements which were made for enforcement. Relatively isolated from the administrative structures charged with the supervision of safety in other sectors of industry, offshore safety became subject to a 'special' relationship. One result was that the industry's self-image of uniqueness was not only reflected but also reinforced by the administrative framework, while another was the emergence of a preoccupation with catastrophe, to the detriment of the more conventional hazards which have arguably been the villain of the offshore piece to date. Most of all, perhaps, the task of promulgating and enforcing safety regulations pertaining to the North Sea was placed in dangerously close administrative propinquity to decision-making processes concerned with other priorities, such as getting the oil ashore as fast as possible (chapter 5).

In chapter 6, I will recount the history of the fraught and ultimately not very successful attempts which were made in the second half of the 1970s to break the effective monopoly held by one Department (the Department of Energy) over the administration of offshore safety. More specifically, I will examine the internal conflict which arose within the state's own bureaucracy in the course of moves to bring the North Sea safety regime into the purview of the more generically oriented system of safety administration developing in Britain at that time under the auspices of the newly established Health and Safety Commission and its executive, the Health and Safety Executive. The very antithesis of special relationships, institutionalized uniqueness and the rest, the intrusion of these organizations into the affairs of the North Sea oil industry touched a number of extremely raw nerves and, not suprisingly, provoked something of a violent backlash from the industry and its former controllers alike. That such should have been the case, I will argue, was the consequence not merely of bureaucratic rivalry but also of other and more deep-seated forces and constraints which have been at work in shaping British policy towards the resources of the Continental Shelf from the very outset; that the outcome of such factors should have been disarticulation and conflict between different segments of the administrative machine underlines the lack of co-ordination and coherent purpose which characterizes the complex workings of the contemporary state.

The last substantive chapter of this book will attempt to construct the profile of offshore enforcement practice. Defining 'enforcement' in broad terms, which include not only the activities of the regulatory agencies involved but also the separate Scottish machinery for the investigation of fatal accidents, this part of the work will concentrate upon some of what I see as the distinctive features of the regulatory regime in action, as well as some of the major problems with which it has had to contend. At one level I will trace the development of 'institutionalized tolerance' with regard to the contravention of offshore safety regulations, a development which has pushed this kind of law breaking a long way into the category of crimes that are called 'conventional' because they are accepted as customary, are only rarely subjected to criminal prosecution and, indeed, are often not regarded as really constituting crimes at all.[19] At another, I will chart the tangled web of chaos and confusion which has come to surround various facets of the way in which the machinery for ensuring safety in the

North Sea operates, and will link this disarray to forces which have a more concrete substance than mere inadvertence or unfamiliarity with a field of safety regulation which is novel. Finally, this chapter will give an account of the far from glorious record of the authorities when they have attempted to bring the full force of the law to bear on offshore violations. While many of the problems, confusions and failings of law enforcement *vis-á-vis* North Sea oil operations find their practical realization in Scotland, their origins, I will suggest, lie geographically and analytically elsewhere.

Thus far is it appropriate for a brief introductory chapter to go in outlining the descriptive content of what is to follow. There is a story to be told, a myth to be dispelled and some censure to be accepted by a nation which has long prided itself on occupying the forefront in the business of ensuring the occupational safety of industrial employees. In the drive towards the realization of one of the 1960's bravest residual hopes, this pride, or at least the entitlement to it, has taken something of a beating. In the rush to get gas and then oil ashore from the North Sea, safety has come a rather poor second. Our historian taking stock might not, after all, be terribly impressed. Nor would he merely be cavilling at a temporary lapse in a long tradition of humanitarian concern. Such concern certainly played its part in the development of onshore safety legislation from the middle of the nineteenth century onwards, but the regulation of working conditions also owes its historical significance to other and less altruistic considerations. Not least, it has arguably played a crucial role in the vital business of legitimizing the social relations of the workplace for more than a century and a half.[20] That such an important task should have taken so long to accomplish in the case of North Sea oil and gas operations suggests that something even more basic may have been afoot in causing the delay and in obstructing the emergence of a coherent and effective system of enforcement.

In asking what this may have been, I am concious of taking a rather old-fashioned view of the sociological imagination as my starting point. So much has happened and so much has been written since C. Wright Mills called for the reconnection of personal troubles to public issues that an appeal to his enjoinder in this respect must seem almost archaic.[21] And yet, I believe, this recipe for social science still offers the best general counsel to any researcher whose work aspires to go beyond detailed description of personal predicaments or social problems and to locate them in terms of the structural forces by which they are so frequently underpinned. Nor is it just that forging links with public issues may stimulate political discussion in areas of social life where, as Weber observed, bureaucracy tends to turn problems of politics into problems of administration.[22] Equally important is the fact that only by making such connections can the sociologist ultimately purport to tell the ordinary man anything that he does not know for himself. Mills, himself, put the point with characteristic eloquence:

> men do not usually define the troubles they endure in terms of historical change and institutional contradiction. The well-being they enjoy, they do not usually impute to the big ups and downs of the societies in which they live. Seldom aware of the intricate connection between the patterns of their own lives and the course of world history, ordinary men do not usually know what this connection means for the kinds of men they are becoming and the kind of history-making in which they might take part. . . . They cannot cope with their personal troubles in such ways as to control the structural transformations that usually lie behind them.[23]

Personal troubles can scarcely come any bigger than being killed on an offshore oil installation. But what are the public issues to which such personal tragedies should be connected, and what are the 'big ups and downs' to which this ultimate lack of well-being should be imputed? As I have already hinted, one of this book's central arguments is that the North Sea's poor safety record has to be viewed against the background of persistent preoccupation with rapid exploitation. At one level, of course, it may be that the industry's own penchant for getting on with things as quickly as possible – for its own sound economic reasons – may be a significant contributory factor in the generation of a high accident rate. More central to my thesis, however, is the argument that the commitment of successive Governments and Departments to the policy of speedy exploration and exploitation permitted a situation to arise in which operations ran on ahead without adequate legislative provision for their safety. Even when such provision was made, the focus on speed still meant that the pace of offshore developments continued to outstrip the formulation of the necessary subordinate regulations. The preoccupation with haste also permeated the practice of enforcement, contributed to the maintenance of a 'privatized' relationship between controllers and controlled, and became one of the main bones of contention in the internecine wrangling which came to surround the administration of offshore safety in the second half of the 1970s. Not least, the urgency with which the objective of getting Britain's oil ashore was pursued left a stream of

legislative and organizational chaos in its wake and thus further
hampered the establishment of effective controls.

These are contentious claims, for which, I hope, subsequent
chapters will provide ample justification. Indictment of speed does not,
however, provide a complete answer to the questions which surround
the connections with which we are here concerned. How can the
recurrent endorsement of speed as the number one priority with regard
to the North Sea be explained? Here, I suggest, the links between
personal troubles and public issues have to be pursued not only up to
but also beyond the totality of the nation state itself. The haste with
which indigenous offshore energy supplies were sought has to be
viewed against the wider backdrop of those 'big ups and downs' of the
world economic system within which oil has come to occupy such an
important place since the Second World War. OPEC's assertion of
strength in the 1970s, with its implications not only for the economies
of the developed countries but also for their previous monopoly on
decision making of any world-scale economic significance, is but the
most obvious example of external factors which must be taken into
account in any attempt to make sense of Britain's oil policy. Even more
basic is the fact that the natural resources of the Continental Shelf first
found their way on to the UK's political agenda at a time when the
post-war US-dominated system of international trade and finance was
about to move into a period of dangerous instability and crisis. And
this, moreover, when the long trajectory of Britain's decline on the
world stage was beginning to dawn on all but the most myopic
adherents to the headier vision of the days of industry and empire. Ere
long, the United Kingdom would have to face up to the economic
discipline imposed by a refurbished system of international finance
which no longer automatically concurred in the view that her aspirations to a major world role warranted special treatment.

These broader developments might seem, in themselves, to comprise
an adequate array of factors to explain why successive governments
would opt to get their hands on the new-found wealth of the Continental Shelf as rapidly as possible and at almost any cost. As what has been
dubbed 'the capitalist world economy'[24] moved further and further
into difficulty and into inflationary crisis, Britain now had a nest-egg
upon which the external exigencies of her predicament could induce
her, if not quite compel her, to draw. Gough puts the general point very
well when he observes in connection with the 'law' of combined and
uneven development that 'nation states are not totally autonomous
entities but elements of this system [the capitalist world system] and

the destiny of any individual country cannot be considered in isolation from developments elsewhere.'[25] In analysing the role which North Sea oil was to be allocated in the shaping of Britain's destiny, discussion of such developments might appear to be almost as sufficient as it is self-evidently relevant.

Movements on a global scale or within the capitalist world system do not, however, allow for the effects of the historically specific and unique features of individual countries at different stages of development. Differentials in such associated matters as levels of class conflict, underlying rates of productivity growth and patterns of dominant interest within the machinery of the state itself all conjoin to ensure that, whatever the overarching forces at work, different countries still follow different paths. As David Purdy observes in this respect, while 'it is a truism that no nation state's development can be understood independently of its international context . . . a *context* is precisely that and not the entire reality'.[26] Stressing the importance of his point for any analysis of the crisis which has overtaken British capitalism since the war, he continues:

> As a component of an international system the nation state has to be accorded relative autonomy. This is because the nation state has been, is and will for the foreseeable future continue to be the major effective locus for the organization and management of economic, political and ideological life. . . . The analysis of British capitalism can never be subsumed under the analysis of the world economy. The task is to locate its 'originality and uniqueness'.[27]

In chapter 4 I will invoke, though not elaborate further upon, the notion of combined and uneven development in order to provide a framework for discussing the political economy of the speed with which the resources of the UK Continental Shelf have been explored and exploited. Charting developments within the international economy and, in particular, within its oil sector, I will attempt to show how these indeed combined with distinctive features of the British economy, class relations and state to generate policies geared to the most rapid possible progress offshore. Inevitably, perhaps, such an undertaking may seem unduly ambitious and possibly even something of an unwarrantable digression for a book which is primarily concerned with safety. In defence, however, I would argue that it is to this concatenation of macro-sociological forces that we must look first for an explanation of the British legal system's poor showing in the context of offshore safety. Viewed from the other and more tragic end of the spectrum, it is only through the prism of such connections that the troubles endured by those who have been killed or injured in the North Sea can be viewed in terms of the 'historical change and institutional contradiction' which earns them anything more than a coincidental place in the 'course of world history'.

Documenting how these macro-sociological forces permeated and not infrequently found conscious expression in such settings as parliamentary debates, Government committees, inter-departmental rows and the routine processes of law enforcement will occupy a substantial proportion of this book. It was in everyday contexts such as these that the constraints which, in my view, were so consequential in shaping the offshore regime had their concrete impact, and it is to this locus that we must turn for evidence if the whole argument is not to be left in suspension at the unsatisfactory level of the *a priori*. Without wishing to enter into complex debates about the degree of autonomy to be accorded to human agency in these matters, I would suggest that documentation of how structures, institutional practices and patterned outcomes are constituted through the medium of human action is a task which sociological accounts can only neglect at their own peril.[28] No violence is done to the internal logic of Marx's famous aphorism if it is inverted to read, 'While men do not make their own history under circumstances chosen by themselves . . . they do nonetheless make it.'[29]

What follows, then, is an attempt at concrete analysis rather than an excursus at the high level of abstraction which some might see the theoretical relationships surrounding the issue of North Sea oil as warranting. Nowhere, for example, will I tackle the thorny issue of economic determinism, whether in the last instance or otherwise, though I suspect that in an area of such crucial and explicit economic significance as oil, such 'determinations' are of more overriding importance than they may be in other contexts. Nor will I attempt to carve out a defensible position for myself in terms of today's highly sophisticated and abstract theoretical debates about the nature of the contemporary state, though again, I am consoled by noticing that a growing number of scholars in this field are acknowledging both the need for concrete analysis and the difficulties which accompany attempts to move from such a high level of abstraction down to the 'lower levels' involved in most research projects.[30] For those whose concern is with such debates, I hope that somewhere in my analysis they will find something of interest and maybe even of more than passing relevance. For the rest, my hopes are twofold; that practition-

ers of the sociology of law will be convinced that there is still space for
the practice of that subject as I conceive it, and that readers in general
may be persuaded to share a little of the sense of shame which I have
come to feel about the other price paid for Britain's oil.

Notes and References

1. E. Hobsbawm, *Industry and Empire*, Harmondsworth, Penguin Books, 1969, p. 321.
2. ibid.
3. 1964, c. 29.
4. For a more detailed discussion of developments in relation to natural gas, see chapter 4 below.
5. *Fuel Policy*. Cmnd. 3438, London, HMSO, 1967, p. 7.
6. Oil developments are more fully discussed in chapter 4 below.
7. Department of Energy, *Production and Reserves of Oil and Gas in the United Kingdom*, London, HMSO, 1974, pp. 1–2.
8. Department of Energy, *Development of the Oil and Gas Resources of the United Kingdom* (hereafter *The Brown Book*), London, HMSO 1981, p. 20.
9. ibid., p. 4.
10. ibid., p. 18.
11. G. Arnold, *Britain's Oil*, London, Hamish Hamilton, p. 100.
12. *Brown Book*, 1981, p. 42.
13. ibid., p. 20.
14. See, for example, *Brown Book*, 1978, p. 27.
15. *Scotsman*, 25 October 1979.
16. Arnold, *Britain's Oil*, p. 175.
17. See chapter 4 below.
18. See, for example, D. I. MacKay, 'North Sea Oil and the Scottish Economy', *North Sea Study Occasional Papers*, University of Aberdeen, 1975, p. 25.
19. For a discussion of 'conventional crime' in relation to the history of onshore factory legislation, see W. G. Carson, 'The Conventionalisation of Early Factory Crime', *International Journal for the Sociology of Law*, vol. 7, no. 1, 1979, pp. 37–60.
20. ibid., p. 46.
21. *The Sociological Imagination*, Oxford, Oxford University Press, 1959, p. 226.
22. R. Bendix, *Max Weber: An Intellectual Portrait*, London Heinemann Educational Books, 1960, p. 433.
23. Mills, *The Sociological Imagination*, p. 3.
24. See, in particular, I. Wallerstein, *The Capitalist World-Economy*, Cambridge, Cambridge University Press, 1979.

25. I. Gough, 'State Expenditure in Advanced Capitalism', *New Left Review*, no. 92, July–August 1975, p. 68.
26. D. Purdy, 'British Capitalism Since the War', *Marxism Today*, vol. 20, September 1976, p. 271.
27. ibid.
28. For an examination of some of the issues involved in what Anthony Giddens calls 'structuration', the dynamic process whereby structures come into being, see A. Giddens, *New Rules of Sociological Method*, London, Hutchinson, 1976, p. 120ff.
29. 'The Eighteenth Brumaire of Louis Bonaparte', *Marx/Engels: Selected Works*, vol. 1, Moscow, Progress Publishers, 1958, p. 247.
30. See, for example, B. Jessop, 'On Recent Theories of Law, the State, and Juridico-Political Ideology', *International Journal of the Sociology of Law*, vol. 8, no. 4, 1980, p. 348. A particularly swingeing attack on 'abstract Marxists' is made by James Petras and his colleagues in their work on Venezuelan oil and the imperial state: 'Abstract Marxism, devoid of operational meaning, leads to a multiplication of taxonomic categories and propositional statements that then are illustrated by an arbitrary selection of historic facts. The lack of scientific rigor in this rather mechanical exercise leads to endless ideological confrontations that, on their own terms, cannot be resolved: it remains a confrontation of positions with the appropriate illustrations. Thus, it falls on empirical analysis to bring out the specific features that identify the imperial state and give meaning to its behaviour. Otherwise reality is stood on its head and abstract theory is considered a way to resolve the imperial-induced crises of our times.' J. Petras *et al.*, *The Nationalisation of Venezuelan Oil*, New York, Praeger, 1977, p. xx. While my own views certainly would not warrant statement in language of anything approaching such strength, I do believe that both the sociology of law and criminology have recently gone through a phase of abstract pre-occupation which, although extremely important and even timely, now needs to be balanced by a resurgence of theoretically informed, empirical research. See also, Erik Olin Wright, *Class, Crisis and State*, London, New Left Books, 1978.

2

Counting the Cost

These new-found resources are conferring enormous benefits on
the United Kingdom. . . . But as with all things there are penalties to
be paid. Too many have already paid the ultimate penalty with
their lives, which is tragically the price so often extracted of
pioneers.
Mr Anthony Wedgwood Benn, MP, 1977

As we saw in the previous chapter, Britain's rise from a position of
chronic energy dependency to join the elite ranks of those nations
which enjoy net self-sufficiency in oil has been astonishingly rapid.
Whether the price that has been paid in terms of death and injury in the
course of this dramatic turn-about was inevitable, however, is a question about which this book canvasses scepticism. On one level, I shall
subsequently argue, the extent to which the 'price extracted' on this
occasion had much to do with pioneering and the attendant hazards of
dangerous frontiers is highly questionable. On another, one of the
work's central themes is that a little less haste might have bought more
time for the development of an adequate system of safety controls to
protect those whose labour would be so vital to the enterprise. As with
the economic implications of North Sea oil, so too with its ramifications for safety; those concerned with the formulation of policy might
usefully have heeded Shakespeare's advice – 'Wisely and slow. They
stumble that run fast.'[1] Before turning to these and other aspects of the
North Sea's chequered history, however, a prior question to be addressed is just how dangerous an undertaking has it been? In short, what
has been the human as opposed to the economic cost?

At several points in the course of this research, I have been warned
about the 'emotive' nature of the terrain into which I was straying in
raising this preliminary question. Not surprisingly, officials involved in
regulating the safety of offshore operations regard the issue as a highly
sensitive one, no doubt in part because any suggestion of an inordinately high casualty rate might reflect badly upon the policy and

practice of the Departments concerned. Furthermore, as will become
apparent in later chapters, the Department carrying the main burden
of responsibility in this context (the Department of Energy) seems to
enjoy a fairly close sense of identity with the industry which it purports
to control and, as a result, reacts badly to any hint that its record is not
a good one. *Mutatis mutandis*, any accusation of high accident rates
touches the industry itself on one of its rawest nerves. Indeed, the issue
is so sensitive that according to the head of one major oil company's
safety department, the United Kingdom Offshore Operator's Association attempts to maintain its own accident statistics 'with a view to
refuting the official picture'. Similarly, one of the papers catering
specifically for oil men pointedly counsels the press to remember that
'more oil men are killed in their cars than in the course of their work'
before it 'goes to town on the casualty rate in the North Sea oil
industry'.[2] The latter's sensitivity on the subject is cogently summed up
in the following extract from an interview with another safety
manager:

> But it's the reaction, you see. For example, there was a coach load of
> people came to grief recently down in England, wasn't there? In fact,
> it was a headline in the paper. But I'm sure that the average public
> reaction to that was, 'Oh dear, that was a disaster!' But these things
> happen on the road, whereas if a diver is killed offshore, then there is
> another immediate outcry about protection of people offshore, and
> what a dangerous industry it is and so on.

In venturing into the arena of such statistics and of offshore risks in
general, one is therefore acutely conscious of entering a highly contentious field in which the problems are not merely technical. Conclusions of the limited kind that may be drawn from the available material
on North Sea accidents are almost bound to aggravate some party to
the controversy and, in so doing, they index the fact that the difficulties
which plague this area are inseparable from the broader pattern of
relationships and processes with which this book is concerned. Statistics,
it should not be forgotten, always carry some political potential, and
North Sea oil is not only a physically volatile substance but also a
politically volatile issue. Added to that, the very deficiencies of the
statistics themselves attest to important features of the North Sea
safety regime. Inadequately co-ordinated with analogous figures for
onshore industries, they point up the fact that, for economic and
political reasons to be discussed in chapters 5 and 6, this sector of
industrial safety legislation has been substantially permitted to plough
its own administrative furrow; incomplete in their coverage, they

reflect the ad hoc nature of a regulatory approach which, always overshadowed by the other urgent considerations associated with North Sea oil, failed to develop a timely, coherent and comprehensive framework for controlling the safety of operations on the Continental Shelf in their entirety. Stated bluntly, many of the problems associated with offshore safety statistics have their provenance in the broader issues of political economy outlined in a subsequent chapter.[3]

So many allusions to what is yet to come might suggest that dealing with statistical patterns at this stage is a somewhat preemptive exercise. Indeed, I have argued with myself at some length as to whether this aspect of the analysis might not be better left to a later point. On balance, however, I have concluded that now is the appropriate time for a brief, prefatory statistical analysis, even if it means commencing substantive discussion with the always complex and, to many people, understandably tiresome business of juggling with figures. For one thing, it is only against the background of the North Sea's safety record, however incomplete, that the full import of another important record can be grasped – the far from creditable history of how UK legislators responded to the evident need for statutory regulation.[4] No less important, limited though they may be, data on offshore accidents may serve to cast some light not only on the question of whether offshore employment deserves its dangerous reputation, but also on some of the patterns of hazard around which a colourful imagery of offshore danger is woven. Thus analysis of this data provides a useful backcloth against which examination of such imagery can subsequently be undertaken.[5] Finally, while I am conscious of the snares of that sociological heresy which accepts official statistics as documenting events rather than the organizational processes involved in their own production, the practical importance of essaying some assessment of offshore risks seems to warrant early consideration of the available data. And for this purpose, the official statistics are all we have to go on. As the statistical appendix to a recent report on offshore safety (the Burgoyne Report) points out, 'from the very practical point of view that these figures are all that are available, they must be considered and conclusions drawn however tentative they may be.'[6]

The Statistical Record

According to the most recent figures published by the Department of Energy, 106 fatalities have occurred on or around installations in the

British sector of the North Sea up to the end of 1980.[7] On top of this, 450 serious accidents had been recorded by the end of the same year. A further 528 'dangerous occurrences' (incidents which could have caused serious injury or actually did involve any of a series of specified eventualities)[8] had been reported since 1974, when relevant figures were first published. To round off the picture, it is necessary to take into account a substantial number of other accidents which, while resulting in three or more days' absence from work, did not meet the Department's criteria of 'serious injury'.[9] Although figures for these are not published by the Department itself, data supplied to the Burgoyne Committee produced a reported total of 3706 such accidents between 1974 and 1978.[10] Figures for the periods before and after this span of five years are not available.

The statistics produced annually in the Department of Energy's *Brown Book*, the source from which most of the above figures are drawn, take cognizance only of incidents taking place on or around installations, because reporting requirements under the Mineral Workings (Offshore Installations) Act 1971 are restricted to installations and a zone of 500 metres (547 yds) around them.[11] Quite apart from any reporting inaccuracies, therefore, they do not represent an exhaustive catalogue of the injuries and dangers which occur in connection with attempts to exploit the energy resources lying beneath the sea bed of the British Continental Shelf. Indeed, as we shall see at a later point, there are several important gaps in the overall statistical coverage of offshore operations. Moreover, the situation is further complicated by the fact that the legal basis upon which the reporting of accidents is mandatory was altered at the end of 1973, with the promulgation of new regulations relating to the notification of casualties.[12] As a result, there has to be some doubt as to whether figures from before and after that regulatory watershed are strictly comparable.

These difficulties are compounded when it comes to the computation of incidence rates as a basis for assessing just how dangerous offshore employment may be. Although accidents which happen on vessels around the installation (i.e., within the 500-metre zone) should, in theory at least, be reported to the Department of Energy, the absence of even any estimates relating to the numbers actually employed on vessels precludes the calculation of a relevant incidence rate. Indeed, when the Marine Division of the Department of Trade, the body responsible for this aspect of offshore safety, was approached for such figures, a representative said that the Department would be grateful if I could tell them how many people are employed in this context.[13] Thus with the exception of diving, for which some broader employment estimates are available, the calculation of incidence rates has to be restricted to those accidents which occur on installations themselves. Even here, the relatively small numbers of estimated employees means that a marginal increase or decrease in accidents during any given year can cause fairly dramatic fluctuations in the incidence rate. For this reason, if for no other, the results of such calculations have to be treated with considerable caution.

Despite all these limitations, the statistics collected by the Department of Energy constitute the only basis upon which we can begin to construct a profile of offshore risks or to identify trends and patterns within them. No less important, such figures represent the only means, other than resort to purely impressionistic accounts, whereby these risks can even be tentatively compared with those encountered in the course of other reputedly dangerous types of employment. Without such comparisons, we have no benchmark against which to gauge the relative extent of offshore risks. Thus while we will return to the deficiencies of North Sea statistics at a later stage, the first task is to establish what they purport to tell us about the danger involved in working offshore.

Table 1 shows the incidence rate for fatalities occurring on or around installations, excluding vessels, from 1969 to 1979. Straightaway the impact of the comparatively small number of men involved in offshore work becomes apparent. During 1972, for example, a reduction of one in the number of deaths, coupled with an increase of less than 600 in the workforce employed on installations, virtually halved the fatality incidence rate. Having said this, however, the overall picture is one which lends some credence to the common belief that, whatever its record in the past, the industry is becoming safer.

One explanation for the evident improvement from 1976 onwards is that North Sea operations become safer as the industry moves from the hazardous phases of exploration, field development and construction into the more stable phase of production itself. Such optimism must, however, be tempered by two considerations. In the first place, while operators claim that they have much better safety control during the production phase,[14] it must be remembered that as time goes on, the amount of maintenance work involving activities such as the use and erection of scaffolds is likely to increase, thus reintroducing construction-type hazards.[15] No less important, the statistical impact which a major production incident could have must always be borne in

TABLE 1
INCIDENCE RATES OF REPORTED FATALITIES ON
OR AROUND INSTALLATIONS (EXCLUDING
VESSELS), 1969–79

Year	No. employed on installations	No. of fatalities	Incidence rate per 1000 at risk
1969	1,450	2	1.38
1970	1,150	1	0.87
1971	1,260	4	3.17
1972	1,850	3	1.62
1973	2,430	2	0.82
1974	4,030	9	2.23
1975	6,300	9	1.43
1976	9,200	16	1.74
1977	12,100	10	0.83
1978	12,500	0	—
1979	10,500	7	0.67

Source: Department of Energy, *Development of the Oil and Gas Resources of the United Kingdom*, 1980, London, HMSO. Crown copyright.

mind. As a draft of a Norwegian report on the overall risk assessment of offshore petroleum activities pointed out with reference to the projected reduction in fatalities as a result of the movement into production:

> the fact that potential hazards within the production phase, fortunately, have not yet led to any multideath accident, call for caution to take this as a true statement. If an accident as the explosion at Flixborough, England, where twenty-eight persons were killed and over a hundred others injured, should occur to a North Sea installation, such an event would dramatically change our figures.[16]

How great is the risk of death in the course of offshore as compared with other types of employment? Table 2 compares the incidence rates for fatalities occurring on installations between 1971 and 1979, again excluding vessels, with those for three of the reputedly more dangerous forms of onshore employment. In order to minimize the effect of annual fluctuations, the table shows average fatality incidence rates for 1971–3, 1974–6 and 1977–9, although in the latter period the lack of onshore statistics going beyond 1977 means that the landward base for comparison comprises only one year. The figures shown in brackets indicate the appropriate factor by which the onshore rates

TABLE 2
AVERAGE FATALITY INCIDENCE RATES, 1971–9

| | *Average incidence rate per 1000 at risk* | | | |
	1971–3	*1974–6*	*(1977)*	*1977–9*
Offshore installations (excluding vessels)	1.87	1.80		0.5
Quarries (including open-cast mining)	0.39 (× 4.8)	0.31 (× 5.8)	0.24 (× 2.1)	
Mines	0.25 (× 7.5)	0.21 (× 8.6)	0.16 (× 3.1)	
Construction	0.20 (× 9.3)	0.16 (× 11.2)	0.14 (× 3.6)	

Source: Health and Safety Executive, *Health and Safety Statistics*, 1975, 1976 and 1977, London, HMSO; Department of Energy, *Development of the Oil and Gas Resources of the United Kingdom*, 1979, London, HMSO. Crown copyright.

would have to be multiplied in order to match their offshore counterparts.

From table 2 it appears that, by the mid-1970s, the likelihood of being killed in the course of employment on offshore installations operating in the British sector of the North Sea had risen to around eleven times that of accidental death in the construction industry, to nearly nine times that of becoming a fatal casualty in mining, and to nearly six times that of being killed as a quarryman. These estimates are in substantial agreement with figures produced by the British Medical Association in 1975, though I have not followed that body's example by pursuing systematic comparisons with the fishing industry or with manufacturing industry in general.[17] While the gap between offshore and onshore rates had narrowed considerably by the last three years of the decade, moreover, it still remained substantial, the relevant factors for arithmetical comparison ranging from approximately 2, in the case of quarrying, to nearly 4 in that of construction. With regard to the other category, mining, the gap appears to be somewhat larger than that shown up by a Norwegian study which puts the overall risk of accidental death offshore at only twice that of the mining industry. However, this difference may be accounted for by the fact that the Norwegian figures are calculated on the basis of thousands of man years worked (arguably a better basis for analysis, but one which cannot be utilized for the British sector) and exclude administrative personnel.[18] Even on their figures, furthermore, offshore oil installations would still seem to be extremely dangerous places to work.

One possible objection to the comparisons shown above is that whereas miners, quarrymen and construction workers leave their place of employment, however hazardous, at the end of the day or shift, workers on rigs and platforms in the North Sea are at risk twenty-four hours per day while offshore. However, examination of accident details included in the Department's quarterly summaries covering 1974–9, the period for which these data have been made available, revealed only three instances in which death might have taken place outside the context of work. One case was simply listed as an 'unexplained fall' and, in the absence of further information, must be listed as possibly having happened during leisure time. A cross-check with fatal accident files revealed that one of the other fatalities of this kind simply involved the deceased's disappearance between shifts, his absence not being noticed for twelve hours because of a recent switch in accommodation. In the remaining case, the victim was descending from temporary sleeping quarters located on top of other temporary accommodation in order to take a shower, when he fell from the top of a vertical ladder which lacked any handrail protection.

Another major objection to such general statistics for offshore fatalities is that they may, of course, be distorted by a small number of relatively major disasters in any given year or, more probably, by the fact that they subsume particularly dangerous occupations. In the first of these contexts, for example, the Norwegian study already referred to notes that twenty-three out of twenty-six fatalities occurring in that sector during 1978 resulted from just two incidents – five from a fire on the Statfjord A platform and eighteen from one helicopter crash.[19] The British sector, however, has witnessed no 'major' accident of this kind since the loss of Sea Gem in 1965, the closest during the period covered by table 2 probably being the drowning of three men in 1977 when two of the victims jumped into the sea in a vain effort to save their colleague (but see note on page 5).

The inclusion of disproportionately hazardous occupations, particularly diving, does produce some distortion in these general estimates, however. In view of this effect, it is therefore all the more unfortunate that although we can calculate and deduct the number of divers killed on installations each year, the statistics do not always permit requisite adjustment to the numbers employed. By extrapolation from the Department's employment estimates, however, this exercise is feasible for the period 1974–6, and the resultant incidence rates for non-diving fatalities per 1000 employees is shown in table 3. Also shown are the less capricious averages for the offshore rates and for those relating to the three onshore occupations already used for purposes of comparison. It is readily apparent that although the exclusion of diving narrows the gap between offshore and onshore somewhat, the North Sea was still maintaining its dangerous lead by a substantial margin.

TABLE 3
INCIDENCE RATES OF NON-DIVING FATALITIES OFFSHORE
(EXCLUDING VESSELS) COMPARED WITH ONSHORE
OCCUPATIONS, 1974–6

	Incidence rate per 1000 at Risk			Average incidence rate per 1000 at Risk, 1974–6
	1974	1975	1976	
Non-diving fatalities on installations	1.6	1.0	1.1	1.23
Quarries (including open-cast mining)				0.31 (× 4.0)
Mining				0.21 (× 5.9)
Construction				0.16 (× 7.7)

Source: Health and Safety Executive, *Health and Safety Statistics*, 1975 and 1976, London, HMSO; Department of Energy, *Development of the Oil and Gas Resources of the United Kingdom*, 1975–7, London, HMSO. Crown copyright.

This brings us to diving itself, by far the most dangerous activity on the Continental Shelf and possibly the most dangerous occupation in Britain. Here, for each year from 1976 to 1978, the Department of Energy's annual *Brown Book* included a separate estimate of the total number of divers working offshore in the British sector (i.e., both on installations and on other locations such as pipelaying barges) to be set against the total number of diving fatalities occurring, irrespective of location. In previous years, the incidence rate of accidental death among divers seems to have been calculated on the basis of the smaller population working and dying on installations. After 1978, and in defiance of the general rule that annual statistics should become more rather than less precise as time goes on, no separate estimate of the numbers of divers employed offshore was given, with the result that the calculation of even a tentative incidence rate becomes impossible. Table 4 shows the high annual incidence rate of diving fatalities between 1974 and 1978, as cited by the Department of Energy itself.

Within the limits imposed by the small numbers involved, particularly in the first two years, these figures speak for themselves. Even at its lowest point in 1978, the fatality rate for diving was more than five

Counting the Cost

times that for the quarrying industry, statistically one of the most
dangerous onshore occupations between 1974 and 1976.[20] If we were
to ignore the different populations and take the average annual fatality
rate among divers across the identical period, the resulting figure of
eight per 1000 at risk would be twenty-six times its equivalent for
quarrying, thirty-eight times the average for mining, and fifty times the
comparable figure for the construction industry. Such an unfavourable
comparison is often explained by the lack of adequate training and
experience said to have characterized North Sea diving until quite

TABLE 4
INCIDENCE RATE OF DIVING FATALITIES, 1974–8

Year	No. employed*	No. of fatalities*	Incidence rate per 1000 at risk*
1974	270	3	11.1
1975	500	3	6.0
1976	1,000	7	7.0
1977	1,200	3	2.5
1978	1,250	2	1.6

*Calculations for 1974 and 1975 are based solely on figures for divers working on
and killed from, installations.
   Source: Department of Energy, *Development of the Oil and Gas Resources of the
United Kingdom*, 1975–9, London, HMSO. Crown copyright.

recently. Whether the pattern of improvement suggested by the figures
for 1977 and 1978 will be sustained in subsequent years is, however,
open to question. By the end of 1979, a further three divers had died in
circumstances falling within the terms of the relevant legislation,
although the lack of figures on employment makes it impossible to say
whether this figure constitutes an improvement or a deterioration in
the situation.[21] Moreover, the continued popularity of conducting
diving operations from dynamically positioned vessels, a system which
uses computer-controlled thruster screws rather than multi-point
moorings to hold the ship on station, gives no particular cause for
optimism in this respect. Indeed, this system has a safety record
sufficiently dubious to have warranted, at one point, consideration of a
total ban by the Department.[22] Whatever the future may hold, however, there is no doubt about the price which Britain's oil has exacted
from divers in the past. Going beyond the period covered in table 4, no

fewer than twenty-three diving deaths have been reported since exploration began. And this figure, as the annual statistics are constantly at pains to underline, does not include fatalities which occur in the course of activities undertaken from pipelaying barges.[23]

Statistics relating to fatalities constitute the best available basis for anything approaching an objective comparison between offshore and onshore risks. When the focus shifts to include accidents which cause serious injury but not death, for example, such comparisons become extremely problematic, since the criteria used for classifying injuries under the Mineral Workings (Offshore Installations) Act differ substantially from those utilized by the Health and Safety Executive onshore. Thus no comparison is possible between what the latter body calls 'Serious (Group 1) Accidents', a category which includes fatalities and a range of other serious injuries, and the combined rate for fatalities and serious injuries occurring offshore. However, the definition of 'serious' used by HM Mines and Quarries Inspectorate is identical to that applying offshore, with the exception that it does not include the same range of injuries arising from electrocution and exposure to radioactive substances.[24] Moreover, although incidence rates covering fatal and serious accidents in these occupations are calculated by the number of man shifts rather than the number of employees, it is possible to reconstruct from various sources an approximate incidence rate based on the latter.[25] Since the Department of Energy's quarterly accident summaries permit us to identify and exclude accidents which would fall outside the definition of 'serious' used for coalmines and quarries, it is therefore possible to arrive at a rough comparison between these two onshore occupations and offshore employment.

Examination of the quarterly summaries provided by the Department of Energy for the period 1974–9 revealed that only five accidents – one in 1975, three in 1978 and one in 1979 – required exclusion under the above procedure. Table 5 shows the resulting incidence for fatal and serious accidents on installations (excluding vessels) during this period, and sets them against the rates for coal mining and quarrying between 1974 and 1977, the latest year for which the requisite statistics are available. What emerges is that during the four years for which direct comparison is possible, the risk of fatal or serious accident confronting the offshore worker peaked at around five times that facing the quarryman and three times that run by the coalminer, falling back to an approximately double risk in both instances by 1977. On the assumption that the onshore figures remained fairly constant in the two years following, it seems plausible to suggest that further converg-ence took place in 1978, but that the gap began to widen once again in 1979. While it should be noted that the coal mining figures used here and elsewhere in this chapter embrace surface as well as underground workers, it is also fair to add that the offshore rates cover some categories of worker which, as we shall see do not figure prominently in the serious accident statistics.

TABLE 5

COMBINED INCIDENCE RATES (PER 1000 EMPLOYEES) FOR FATAL AND SERIOUS ACCIDENTS 1974–9 (EXCLUDING VESSELS)

	1974	*1975*	*1976*	*1977*	*1978*	*1979*
Offshore	6.9	8.6	7.2	3.7	2.4	4.3
Quarrying	2.0 (× 3.4)	1.6 (× 5.4)	1.8 (× 4.0)	1.9 (× 1.9)	—	—
Coal mining	2.1 (× 3.3)	2.5 (× 3.4)	2.3 (× 3.1)	2.1 (× 1.8)	—	—

Source: Health and Safety Executive, *Health and Safety Statistics*, 1975–7, London, HMSO; Department of Energy, *Development of the Oil and Gas Resources of the United Kingdom*, 1974–80, London, HMSO. Crown copyright.

When we move from fatal and serious accidents to another report-able category, those which cause three or more day's disablement from work but are not classified as serious, comparisons between onshore and offshore records become even more difficult. Reporting standards vary widely from sector to sector, even onshore, and the results of such comparisons could therefore be extremely misleading. Thus, for exam-ple, the statistical appendix to the Burgoyne Report estimated the incidence rate per 1000 at risk for *all* reported offshore accidents (i.e., a catch-all category dominated numerically by these less serious in-juries) at around eighty for 1976, dropping to seventy in 1977 and rising again to 107 in the following year.[26] Against this, it noted, the comparable figures for mining hovered around the 200 mark from 1972 to 1976. On the basis of this not unflattering difference, the commentary felt able to conclude that overall the offshore worker is 'about half as likely as a miner' to meet with a reportable accident of some kind.[27] Unfortunately, however, the compilers of this particular document omitted to add that the system by which 'three-day' acci-dents in the mining industry are notified to the authorities not only is completely different, but also differs in such a way as to encourage a very high rate of reporting.[28] Nor, it should be added, can we be entirely confident that the standard of reporting in the offshore oil industry itself is a particularly high one when it comes to less serious accidents.[29]

What we are left with, then, is a partial picture based on the risk of death or serious injury among workers on installations and of death among the more broadly based group of divers. Incomplete as it may be, however, the picture is one which suggests that, at least until very recently, the dangerous reputation of the North Sea has been fully earned. Indeed, since the offshore employment figures used in the

TABLE 6

INCIDENCE RATES OF REPORTED FATAL AND SERIOUS ACCIDENTS ON OR AROUND INSTALLATIONS (EXCLUDING VESSELS), 1968–79

Year	No. employed on installations	No. of fatalities	No. of serious accidents	Annual combined incidence rate
1968	1,210	3	21	19.8
1969	1,450	2	19	14.5
1970	1,150	1	12	11.3
1971	1,260	4	15	15.1
1972	1,850	3	17	10.8
1973	2,430	2	22	9.9
1974	4,030	9	19	6.9
1975	6,300	9	46	8.7
1976	9,200	16	50	7.2
1977	12,100	10	35	3.7
1978	12,500	0	33	2.6
1979	10,500	7	39	4.4

Source: Department of Energy, *Development of the Oil and Gas Resources of the United Kingdom*, 1980, London, HMSO. Crown copyright.

above calculations are based on surveys apparently carried out during or near to the *peak* period of activity each year,[30] there is a possibility that the dangers, as measured by comparative incidence rates, are under- rather than overestimated. To be sure, as can be seen from table 6,[31] the record improved more or less steadily as the 1970s progressed, every year but three showing a decline in the combined incidence rate for fatal and serious accidents over the one preceding it. Moreover, it can justifiably be claimed that even if we were to insist on treating the 1979 figures as more than a statistical hiccup, which is all that they probably are, the combined incidence rate has been cut by something approaching five times in the space of a dozen years. On the other hand, however, the very fact of such a reduction only serves to underline just how much room there was for improvement during the greater part of the period with which this book is concerned.

The undifferentiated inclusion of particularly hazardous activities, like diving, in tables 5 and 6 once again raises the question of how these injuries and risks are distributed between different facets of the offshore operation. While this problem cannot be satisfactorily resolved in terms of incidence rates, a broad indication of the activities giving rise to most accidents, irrespective of the numbers employed, can be obtained from examination of published statistics showing accidents according to the activity involved. Table 7 gives such a distribution for the fatal and serious accidents which were reported between 1969 and 1979. Since we are not concerned with employment figures at this point, injuries occurring on vessels are included.

TABLE 7
SERIOUS AND FATAL ACCIDENTS BY ACTIVITY,
1969–79 (INCLUDING VESSELS)

Activity	No. of fatal accidents	No. of serious accidents	Total	%
Drilling	16	142	158	37.6
Cranes	10	47	57	13.6
Maintenance	7	36	43	10.2
Construction	11	31	42	10.0
Diving	22	18	40	9.5
Vessels	10	30	40	9.5
Production	0	17	17	4.0
Helicopters	2	4	6	1.4
Domestic	0	1	1	0.2
Unallocated	0	16	16	3.8
Totals	78	342	420	99.8

Source: Department of Energy, *Development of the Oil and Gas Resources of the United Kingdom*, 1979 and 1980, London, HMSO. Crown copyright.

What emerges here is a picture dominated by accidents associated with drilling, followed at some distance by injuries resulting from crane operations. Thereafter, accidents in the course of maintenance and construction work each contribute around 10 per cent to the total – in fact, the former made quite a dramatic jump from five to thirteen serious injuries between 1978 and 1979 and thereby possibly reflected the increasing amount of such activity referred to earlier. Vessel and diving accidents follow closely behind, the particularly dangerous nature of the latter occupation being reflected in the fact that it both leads the fatality league table and, alone, has more deaths than serious injuries recorded against it. Production and helicopter activities rank as relatively insignificant sources of accidents causing death or serious injury, though in the second of these contexts the British record differs sharply from that of the Norwegians.[32] Even with regard to the British sector of the North Sea, moreover, it should be noted that one of the two fatalities associated with helicopters resulted from an incident in which a Sikorsky S58T carrying ten people spun off the helideck of a fixed platform on to the deck of a derrick barge 140 feet below, narrowly missing the divers' support system on that vessel before catching fire. In the circumstances, the casualty list of one dead and four seriously injured was remarkably low (but see note on page 5). No such qualification need be made with regard to the 'domestic' category, however, workers like cleaners, caterers and wireless operators appearing to be relatively safe from the risk of major accident. Employees in this category accounted for some 15 per cent of the offshore workforce in 1979.[33]

The above pattern alters somewhat when another genre of accident, that which might, but fortunately does not, result in loss of life or serious injury is taken into account (table 8). In that context, for example, crane and vessel operations displace drilling as the major sources of such 'dangerous occurrences', while production moves into

TABLE 8
DANGEROUS OCCURRENCES BY ACTIVITY, 1974–9

Activity	1974	1975	1976	1977	1978	1979	Total	%
Cranes	10	25	21	18	30	20	124	30.2
Vessels	10	7	12	12	13	16	70	17.1
Drilling	8	5	13	6	9	13	54	13.2
Production	—	4	9	14	15	7	49	11.9
Maintenance	4	—	5	6	7	23	45	11.0
Construction	—	5	6	7	9	3	30	7.3
Diving	2	3	3	1	3	8	20	4.9
Domestic	2	2	1	2	2	2	11	2.7
Helicopters	—	2	1	2	1	1	7	1.7
Totals	36	53	71	68	89	93	410	100.0

Source: Department of Energy, *Development of the Oil and Gas Resources of the United Kingdom*, 1977–80, London, HMSO. Crown copyright.

fourth place, ahead of maintenance and construction, another reminder that we should not be too complacent about the diminution of
risks predicted for this phase. So too with maintenance; the sharp
increase in dangerous occurrences involving this kind of activity during 1979 is evident. More generally, there is a notable increase in the
number of such occurrences over the six-year period since records
were first published, though this is often attributed to an improvement
in standards of reporting.

Statistical Cautions

Thus far in this chapter I have reviewed the official record pertaining to
offshore accidents, set it against those for some other dangerous
occupations and constructed a very broad profile of the main activities
which entail high risks for oil workers in the North Sea.[34] Overall, the
picture which emerges is one suggesting that up until the late 1970s,
employment on installations in the British sector was very hazardous
indeed. Thereafter, the risks diminished quite substantially, but still
remained high compared with those to be encountered in occupations
such as quarrying and mining, at least as far as death or serious injury
are concerned. As for the pattern of activities involving greater danger,
the prominence of operations such as diving and drilling would seem
to attest to the fact that some of these risks are a function of fairly
unique hazards, while the macabre contribution made by activities like
construction, maintenance and craneage would appear, at first sight,
to bear out the common belief that many of the North Sea's dangers
stem from the exigencies associated with the conduct of conventional
industrial operations in a uniquely hostile setting.

To these latter issues we shall be returning at some length, and with
some scepticism, in chapter 3. Here, however, a necessarily prior issue
is that of the reliability and comprehensiveness of the official figures
upon which the above discussion of the offshore safety record is based.
In this context, of course, the problems are by no means unique to the
North Sea. At one level, during recent years it has become something of
a welcome orthodoxy among students of law enforcement in general
to assert that the statistics produced by particular agencies frequently
tell us more about the social processes involved in their production
than about the phenomena they purport to measure. At another, and
one that is closer to the subject in hand, it is also frequently acknowledged that the reporting of accidents which take place onshore is,

at least in the case of some industries, substantially less than complete. A Health and Safety Executive report published in 1980 underlined the problem:

> there are wide variations in the level of accident reporting. It is estimated that while about 75 per cent of reportable accidents in manufacturing establishments subject to the Factories Act are reported, the proportion in the construction industry is only about 50 per cent and in premises subject to OSRP [Offices, Shops and Railway Premises] Act it may be as low as 25 per cent. But reporting in NCB [National Coal Board] mines and under the Railways Acts approaches 100 per cent.[35]

As far as North Sea oil operations are concerned, one thing is certain straightaway – there are some glaring statistical gaps in the overall coverage of accidents which take place in the course of such activities undertaken in the British sector. Even in the extreme context of fatalities, for example, the total of 106 which was given earlier does not cover all of the deaths which have occurred offshore in the course of such operations. Between 1974 and 1979, at least ten other workers were killed, excluding a number who died offshore from natural causes. Diving connected with pipelines located on the Continental Shelf accounted for four of these fatalities, while another two divers were killed while working on pipelaying operations within territorial waters. Accidents on vessels and barges working outwith the 500-metre zone caused the deaths of three other workers, while the final fatality occurred on a rig which was 'stacked' in the Firth of Forth.[36] Nor, as we shall see, can this list be accepted with any great confidence as exhaustive.

The purpose of drawing attention to these additional deaths is not primarily to demonstrate that the Department of Energy figures used in the earlier analysis stand in need of substantial upward revision. Indeed, the above examples were deliberatly taken from that Department's annual reports, which frequently allude to other deaths which fall outside the terms of its own legislation and are not therefore included in its tabular presentation of accident statistics. Rather, the purpose is to underline the extraordinary difficulty which attends any attempt to reach definitive conclusions about the total rate of death, or, even more, of other injuries sustained in the course of offshore employment. Nor is the British sector of the North Sea alone in this respect. Similar problems arise in connection with offshore activities in the United States, where data on the total number of accidents occurring on the

Outer Continental Shelf are collected, but the absence of any employment figures precludes any estimate of incidence rates.[37] Indeed, at one point in the mid-1970s the US Occupational Safety and Health Administration was reduced to extrapolation from British research in order to assess the death-risk faced by American divers![38] Closer to home, the Norwegian authorities admit that accidents taking place outside the 500-metre zone have not always been officially reported.[39] As the International Labour Office reported in 1978, such gaps can arise out of 'a jurisdictional vacuum . . . where neither the nation in the name of which the installation is registered, nor the nation in whose continental shelf area it is working takes responsibility'.[40] More generally, the same body could not forbear to comment somewhat cuttingly on the 'evident' fact that 'the "state of the art" of occupational safety and health statistics for the offshore petroleum industry demonstrates the need for greater efforts in the development of improved statistics at both national and international levels.'[41]

Some of the jurisdictional issues underlying the above problems will be discussed at a later stage.[42] Here, however, it is relevant to add that in the British context an attempt has been made to fill one of the most obvious lacunae, that relating to pipelaying barges operating outwith the 500-metre zone embraced by reporting requirements under the Mineral Workings (Offshore Installations) Act 1971. Subsequent to the passage of the Petroleum and Submarine Pipelines Act 1975,[43] regulations requiring notification of a variety of occurrences taking place in the course of pipelaying activities were promulgated in 1977.[44] Under these regulations, accidents were to be reported to the Department of Energy, but the situation became complicated shortly thereafter by the fact that responsibility for occupational safety on such barges was taken over by the Health and Safety Executive.[45] According to evidence given to the Burgoyne Committee by Shell UK Exploration and Production, the system which subsequently emerged was that accident details (apart from diving) were passed on to the Executive by the Department of Energy, while all emergencies had to be notified directly to the former body by telephone.[46] Parenthetically, it is worth adding here that such arrangements possibly epitomize the chaos which was to characterize the offshore safety regime, to which we shall be returning at a number of subsequent points. Chaos apart, however, members of the Executive remained sceptical for a considerable time as to whether they were receiving the requisite data from whatever source. As late as 1979, one senior official explained in an interview that the Executive only really got to hear about fatalities, that this was usually through the police and that, even then, he could not be entirely certain of being informed about deaths occurring on foreign-registered barges. One of his colleagues openly admitted to a feeling of helplessness:

> These are reported to us on a sort of courtesy basis. . . . I say to whoever, 'It would be nice to know what accidents you are having. You know, there's an enforceable procedure under the Factories Act, you know, do you think we could go along with it?' So it's a question of whether they do so. I don't know how soon he's going to do it; I really have no idea. I've only had one from X and there's not been a dickie bird from anywhere else. This maybe indicates they are being terribly safe or it maybe indicates they are not reporting the accidents. I don't know, and really I have no way of finding out other than asking them.

If pipelaying accidents are not a complete statistical blank, the same cannot be said for injuries sustained in the course of some other activities outside the 500-metre zone. While those occurring on British-registered vessels in such a location should be notified to the Department of Trade under the Merchant Shipping Acts,[47] that organization does not maintain separate statistics for merchant shipping accidents connected with offshore oil operations. In consequence, it is impossible to include even those casualties occurring on British vessels outside the zone in any statistical profile. As for foreign-registered vessels, they are not obliged to report their accidents to the British authorities. All in all, the unsatisfactory situation *vis-à-vis* casualty statistics pertaining to designated areas of the Continental Shelf was summed up, perhaps even then a little optimistically, by the Department of Energy's *Brown Book* for 1978:

> Accidents on or around rigs and platforms covered by the Mineral Workings (Offshore Installations) Act 1971 must be reported to the Department of Energy. In addition regulations introduced in 1977 under the Petroleum and Submarine Pipelines Act 1975 now require the reporting of accidents in the course of work on pipelines. There are some foreign-registered vessels such as crane barges and derrick barges on which reliable information would be impossible to collect, and in connection with which there is no obligation to report accidents. Finally, deaths on British-registered vessels must be reported to the Department of Trade, although again reliable information on foreign registered vessels, including accidents associated with them, is not available.[48]

When the focus shifts to the narrower range of accidents occurring on or around installations themselves, it is pretty generally agreed that, once again, the figures for reported fatalities are accurate. As we move down the scale of seriousness, however, this agreement disappears. On the one hand, the Department of Energy insists that its records are substantially correct, with the exception of those for minor accidents on vessels which should be, but normally are not, reported to the installation manager. According to one senior official, there would be no advantage to be gained from concealing information about serious accidents, not least because 'there are plenty of people on these platforms who are only too willing to make accusations about failing to report – anonymous letters and this sort of deal.' Nor, it is claimed, is there much likelihood of such accidents being systematically reported as minor in nature, since this kind of inconsistency would be 'picked up in the office'. In general, the Department's confidence with respect to the accuracy of its figures for fatal and serious accidents was echoed by the Burgoyne Committee, which concluded that it was 'most unlikely' that such events would go unreported.[49] As for minor accidents, the official view is that these are, if anything, over-reported, partly through the over-zealous recording of things like bruises and splinters in the eye, and partly because a very minor accident can easily qualify as disabling a worker for three days or more by the time the injured person has been helicoptered ashore for medical examination. Indeed, the Health and Safety statistics, which include a total figure for all accidents reported under the Mineral Workings (Offshore Installations) Act, go so far as to say that 'if an employee goes ashore for a medical check-up or X-ray following an accident, he will virtually automatically be recorded as having been absent from work for more than three days.'[50]

Against this view, on the other hand, we have to place other opinions encountered in the course of this project. Thus, for example, the safety manager for one oil major readily conceded that the official statistics are possibly 'a little of an underestimate of the real situation', though this did not lead him to modify his view that the press reaction is still 'out of proportion'. More forcefully, several offshore workers described a general attitude to accidents that would seem conducive to less than accurate reporting. Sometimes, indeed, the alleged discrepancies were acute, to say the least. According to one roughneck, injuries such as the loss of fingers or crushed feet are likely to be called 'minor', although he could not confirm whether this is how they would be formally recorded. Another ex-rig worker reported the practice on

an American installation as being to 'hush up' accidents, adding that
medical personnel try to avoid reporting some accidents because they
think such reports may reflect badly on themselves. A catering worker
recalled how a colleague whose foot had been 'crushed' was told that
he could have an inside job, like painting, but could not be flown
ashore, while another injured roughneck had allegedly fared even
worse:

> I can remember – this was on X – that someone got his hands crushed
> and he got a couple of nails torn out. And he said, 'I want to go
> ashore', and the guy said that when he was doing the job he lost nails
> on all five fingers and never even stopped to take a breath, 'So you get
> back to work.' That was his attitude. So I think this is one of the
> things that is worth investigating. By their standards this was a minor
> accident; this was one that doesn't come into the statistics. So when
> the Department say there is nothing wrong, it's not a deliberate
> whitewash on their part, because they are talking about things which
> haven't even been reported to them.

Real or feigned, such an uncompromising endorsement of the work
ethic is not, however, the only factor which could conceivably affect
the likelihood of accidents being properly reported. In addition, for
example, the use of incentive schemes, whereby groups of workers are
given safety awards, provided that the entire group has a record free of
lost-time accidents over a specified period of time, could have a detrimental effect, particularly on the reporting of less serious accidents.
Watches, lighters, anoraks and even silver tea services can be won
under this system, described in one press report as 'a cross between
Double Your Money and Green Shield Stamps', though it should be
added that I came across only one instance in which it was claimed that
these latter were literally being used in this way.[51] While it is difficult to
assess what impact such systems may have on official reporting
behaviour, there is little doubt that they can provide an incentive for
the minimization of injuries which have occurred. On one installation
visited during the research the present writer witnessed a safety officer
reassuring a squad of highly concerned roustabouts that they would
not forfeit their entitlement to the anorak being shown around simply
because the crane driver had lost the top of one of his fingers a few days
earlier! While the immediate incentive to minimization in an instance
such as this may revolve around the material aspirations of the workforce, other and no less material considerations have been suggested as
providing management with a motive for playing down the seriousness
of accidents:

> Both the Norwegians and the British also accept that the offshore
> industry is a tough world where many personal injuries will be
> treated far more lightly than would be the case on land. . . . It has also
> been suggested that many operators are reluctant to accept imposed
> stoppages that could result in lost downtime or open themselves to
> official probes. Existing regulations call for a halt to work following
> serious incidents to allow time for investigations to be carried out.[52]

Neither the use of safety incentive schemes nor some measure of
reluctance to report incidents which may spark off an official investigation is, of course, the exclusive prerogative of the North Sea oil
industry. Particularly where an incident of the latter kind has left no
serious injury in its wake, the temptation to keep the relevant information confidential is perhaps almost understandable. With the remote
geographical location of North Sea oil operations, however, not to
mention the industry's penchant for secrecy,[53] the chances of doing so
successfully might seem to be considerably enhanced. Thus it comes as
no surprise to discover that it is in the area of 'dangerous occurrences'
that the greatest doubts have been cast upon the reliability of the
published statistics.

In this context, the Department of Energy itself acknowledges having had considerable difficulty in the past and, indeed, at one point
even contemplated a prosecution *pour encourager les autres* in this
respect. As the then Head of Operations and Safety wrote in connection with the case in question,

> We have been preaching this gospel to all the companies and have
> not been one hundred per cent successful in getting them to report
> such accidents as crane boom collapses (which would certainly kill
> anyone in the way) or the virtual explosion of a pump at extremely
> high pressure where, by the grace of God, again nobody was hurt.

While the more recent increase in the number of dangerous occurrences reported may, as the authorities suspect, be partly due to
improved standards in this context, the implication for the accuracy of
the earlier records is not entirely flattering. Nor is it by any means
certain that the Department's evangelism has yet achieved total conversion. A trainee production supervisor, for example, recalled how he
had been openly introduced to the principle that you do not, if possible, report particularly untoward occurrences taking place within the
production process: 'If you do, you will only have a whole crowd of
people who know nothing about it crawling all over you.' Another

example, this time of a dangerous occurrence being down-graded, was recounted by a union official in 1978:

> I had a fellow in here this morning – I won't mention what job he's got or who he works for – but there was a fire out on one of these platforms, and it took place among oxygen and acetylene cylinders, and it was hushed up by the company to the extent that it was put down as a very minor fire. But in fact it was not a minor fire at all; two or three minutes later you would have had oxygen and acetylene cylinders flying all over the place. This fellow actually had to write a report, but then he was asked to tone it down, and that was the report that went to the Department.... This fellow now finds himself in difficulty of being sacked because of the fire, which is another matter altogether. But as I said, 'You have now become a party to the company's collusion . . . you actually signed the report that went to the Ministry.'

Particularly worrying in this connection, and in much more than the statistical sense, is what seems to be a pronounced tendency to neglect the reporting of structural and other failures which could potentially have been disastrous. Indeed, according to one Certifying Authority, such failures are not generally notified *unless* they involve some more or less catostrophic consequences which are unlikely to escape public attention. Given what at this point can only be referred to as the possibility that welding defects played some role in precipitating the Alexander Kielland disaster, moreover, it is doubly disconcerting that the reporting of dangerous occurrences of this kind appears to be fairly minimal. According to the Burgoyne Report, 'few if any' of eleven weld failures which had been recorded by Lloyd's Register in the course of surveying sixty-seven installations had subsequently surfaced in the Department of Energy's records.[54]Furthermore, while none of these failures had actually caused accidents, the Report did not hide its concern about the potentially disastrous implications:

> It is interesting that weld failure has been identified as the cause of only five accidents in nearly six years. . . . The incidence of weld failures is nevertheless known to be quite large from information scattered in the files, and it is fortunate that so few have caused accidents. That they are not considered – and reported – as dangerous occurrences is perhaps less fortunate. It is known that in a number of semi-submersibles, weld failures have been so extensive that the safe operation of the entire rig was in jeopardy.[55]

The work carried out in the course of the Burgoyne investigation suggests, then, that the official statistics on dangerous occurrences may represent a dangerous underestimate of the real situation which exists in the North Sea. As far as other accidents, and particularly non-fatal ones, are concerned, however, we can only go along with the same Committee's view that allegations of under-reporting cannot be confirmed or disputed 'without exhaustive inquiries'.[56] Nor has it been possible to conduct inquiries on such a scale within the scope of the present investigation. Whereas, for example, Department of Health and Social Security statistics on industrial injury claims permit a rough cross-check to be made on the reliability of accident reporting in other industrial sectors, the breadth of the data categories which are used in this context does not allow a similar exercise to be carried out for the numerically small, if economically vital, offshore oil industry.[57] High rates of staff turnover, particulary in the lower-status occupations, substantial numbers of foreign employees and the almost diasporic way in which rig workers disperse upon coming ashore all combine to place obvious obstacles in the path of other and more ad hoc research techniques. Indeed, one conclusion that can be stated even thus early in this report is that there is a pressing need for such an investigation to be mounted by those authorities which could marshall the requisite expertise and data.[58] That such has not already been done is perhaps in itself an indication of the low priority which has been allocated to safety in the rush to get Britain's oil ashore.

In the meantime, what we are left with is two conflicting viewpoints on the reliability of offshore accident statistics. According to one interpretation, the most that can be said is that there must remain a suspicion that the official figures understate the situation – that offshore employment deserves an even more dangerous reputation than it has already acquired. According to the other view, such suspicions are groundless and only attain the semblance of credibility as a consequence of fragmentary allegations which, as one Department of Energy official claimed in interview, almost invariably turn out to be unjustified when they are followed up. What cannot be gainsaid, however, is that even if the figures presented in the first part of this chapter are accepted as accurate, they suggest that the North Sea bonanza has been very costly in terms of death and serious injury. Furthermore, the gap which existed for so long, and which to some extent still exists, between offshore employment and dangerous onshore occupations in terms of casualty rates is something that requires explanation. Whether or not frontier technology and uniquely adverse operating conditions provide an adequate answer in this respect is the issue to which the next chapter is devoted.

Notes and References

1. *Romeo and Juliet,* Act II, scene 3.
2. *Oilman,* 11 August 1979.
3. See chapter 4 below.
4. See chapter 5 below.
5. See chapter 3 below.
6. *Offshore Safety,* Cmmd. 7841 (henceforward, the Burgoyne Report), London, HMSO, 1980, p. 104.
7. *Development of the Oil and Gas Resources of the United Kingdom,* (henceforward *The Brown Book*), London, HMSO, 1981, p. 44.
8. The list of such occurrences requiring notification is given in Form OIR/9 and is divided into those that could have directly caused serious bodily injury and those involving the integrity of the structure of the installation, attendant vessels or aircraft in question.
9. Listed on Form OIR/9.
10. Burgoyne Report, p. 107.
11. 1971, c. 61. The relevant regulations are the Offshore Installations (Inspectors and Casualties) Regulations 1973, SI 1973, 1842.
12. Prior to these regulations accidents in 'the licensed area' were reportable under licence terms.
13. *Brown Book,* 1981, p. 13, explains that estimates of the numbers employed on vessels and in other offshore operations associated with exploration and exploitation are now available. Unfortunately, however, the result has been that now only a total figure for offshore employment is given. This means that the calculation of incidence rates, even for installations themselves, is no longer possible from the published statistics, which seem to be becoming more, rather than less, opaque as time goes on.
14. *Safety Offshore: Risk Assessment* (henceforward, *Risk Assessment*), Royal Norwegian Council for Scientific and Industrial Research, Report No. 26–7/2, Trondheim, 1979, p. 76.
15. ibid., p. 124; see also *Brown Book,* 1980, p. 15.
16. *Risk Assessment,* Report 26–7/1, Draft 2, 1978, p. 65.
17. British Medical Association Scottish Council, *The Medical Implications of Oil Related Industry,* Edinburgh, British Medical Association, 1975, p. 4. The report states that for oil rig workers, the annual average fatality rate at that time appeared to be 'double the fatality rate for deep sea trawlers in near and middle waters, 10 times the fatality rate in coal mining, more than 10 times the rate in the construction industry and 50 times that of factories in general'. In the statistics given in the present instance, quarrying has been included because, according to the Health and Safety Executive, it is 'consistently the most dangerous inland industry'. Health and Safety Executive, *Health and Safety Statistics 1977,* London, HMSO, 1980.
18. *Risk Assessment,* Report No. 26–7/2, 1979, p. v.

19. ibid., p. 73.
20. See p. 21 above.
21. *Brown Book,* 1980, p. 45.
22. *Oilman,* 16 June 1979.
23. See, for example, *Brown Book,* 1981, p. 44, appendix 10, note 1.
24. The definition of serious (reportable) injuries utilized by HM Mines and
     Quarries Inspectorate are given in *Health and Safety Statistics, 1977,*
     London, HMSO, 1980, p. 33. The definition used in connection with
     offshore accidents is given on Form OIR/9, the form on which injuries
     have to be reported.
25. The *Health and Safety Statistics* give the total number of accidents
     reported from quarries and mines and an incidence rate per 100,000
     employees for all reported accidents. Using these figures, it is possible to
     calculate the number of employees taken as being at risk in the compila-
     tion of the relevant statistics. Since the same publication also provides
     figures for the number of fatal and serious accidents reported, an
     incidence rate for these can then be calculated. A separate table showing
     incidence rates for fatalities alone, is also provided, and, using the
     figures for numbers employed (as calculated by the above method) and
     those for the total number of fatalities (as published), it becomes
     possible to check on the employment figures which have been used. No
     discrepancy of more than 0.1 per 100,000 was encountered in the
     fatality incidence rates calculated by this method and those published
     by the Health and Safety Executive.
26. Burgoyne Report, p. 109.
27. ibid., p. 104.
28. Accidents giving rise to disablement for more than three days are
     not reportable unless they qualify as serious by the specified criteria. But
     the National Coal Board, which employs around 99 per cent of the
     labour force, provides the Inspectorate with its own figures. *Health and
     Safety Statistics, 1977,* p. 33. The NCB figures are, however, compiled
     with the assistance of Department of Health and Social Security data on
     industrial injury claims. As a consequence, the level of reporting of such
     mining accidents is reckoned to be almost 100 per cent.
29. See p. 34 ff. below.
30. Up until 1980, the Department of Energy's employment figures were
     based on a survey carried out on one or two dates in mid-summer or
     early autumn each year.
31. The small variation between the incidence rates shown here and those
     given in table 5 is accounted for by the fact that the accidents excluded
     in the former instance because they did not fit the criteria of seriousness
     used by HM Mines and Quarries Inspectorate are included here.
32. See p. 22 above.
33. *Brown Book,* 1980, p. 44.
34. For an attempt at a more detailed analysis of the circumstances involved
     in mishaps, see Burgoyne Report, pp. 116–17.

35. *Health and Safety Statistics, 1977*, p. 3.
36. See *Brown Books, 1975–80*.
37. International Labour Office, *Safety Problems in the Offshore Petroleum Industry* (henceforward, ILO, 1978), Geneva, International Labour Organization, 1978, p. 45. See also Burgoyne Report, p. 105.
38. ILO, 1978, p. 48.
39. ibid., p. 41.
40. ibid., p. 39. The use of the term 'installation' here is misleading, since any structure classified as such would, at least in the British case, be covered by the relevant regulations. However, the ILO does not appear to be alone in slipping into such ambiguity. The Department of Energy statistics on employment for 1980 state that 22,000 workers are employed on 'installations', but the text explains that this figure includes those working on 'service vessels, survey teams, construction support barges etc.'. *Brown Book, 1981*. pp. 13 and 44.
41. ILO, 1978, p. 39.
42. See chapter 7 below.
43. 1975, c. 74.
44. SI, 1977, 835.
45. See chapter 6 below.
46. Burgoyne Report, p. 137.
47. 1970, c. 36; 1974, c. 43.
48. *Brown Book, 1978*, p. 14.
49. Burgoyne Report, p. 43.
50. *Health and Safety Statistics, 1976*, p. 5.
51. *Oilman*, 25 February 1978.
52. ibid., 4 March 1978.
53. See chapter 5 below.
54. Burgoyne Report, p. 114.
55. ibid.
56. ibid., p. 43.
57. It is this method of cross-checking, among other things, that allows the Health and Safety Executive to be fairly confident of the high level of reporting from mines and, conversely, to put that for the construction industry at around 50 per cent. *Health and Safety Statistics, 1977*, p. 3.
58. Such an investigation was called for by the Trades Union Congress in its evidence to Burgoyne. Burgoyne Report, p. 289.

3

Images of Danger

Anyone who has ever ventured out into the North Sea off the Scottish coast knows it is just as wild a setting as India or Africa ever was in the empire days. Surely now, ten years after oil was first discovered there, it should have found its Kipling or Conrad to describe the incredible daring of bringing oil up from the sea-bed through the constant high swells and the sudden violent storms of this most untamed and unpredictable of environments. There certainly have been enough adventures to satisfy any chronicler — and enough violent deaths.
'Empire and after — or plain tales from the rigs',
Guardian, 1979

Frontiers and Frontiersmen

When the events lying behind the statistics so starkly presented in the last chapter actually take place, they sometimes make excellent headlines. Quite apart from the more spectacular catastrophies, such as the loss of Sea Gem in 1965, the blow-out on the Ekofisk Bravo platform in 1977 or the more recent Alexander Kielland disaster, offshore accidents can occasionally generate a considerable amount of media attention. To take an example to which we shall be returning in some detail at a later stage, the deaths of two divers in the Thistle field on 8 August 1979 achieved scarcely less headline prominence than the tragedy, a few days later, in which a single yacht race saw the loss of more people than had been killed on or around British offshore oil installations during the entirety of the previous year.[1] To the popular press, the fact that the story involved a diving bell which had become separated from its mother ship some 500 feet above, suggested obvious and dramatic captions. 'TRAPPED IN STEEL TOMB' and 'SEA TRAP' monopolized the front pages of the *Daily Record* and *Scottish Daily Express* respectively, though the effect was possibly somewhat diminished in the latter instance by an unfortunately inaccurate sub-headline proclaiming 'Dive-bell men safe after 18 hours below'. In the less sensational press, both the *Glasgow Herald* and the *Scotsman* carried seven-column front-page headlines referring to the tragedy, while the story was given the lead in the *Guardian* under the heading 'Trapped divers killed in North Sea bell'. Even the *Daily Telegraph*, though cautiously awaiting word of what would be found now that the bell had been recovered to the surface, carried an account of the '15-hour ordeal for bell divers' on its front page.[2]

That an event such as this should receive such coverage, even though it is matched in terms of tragedy and outstripped in statistical terms by the daily carnage on the roads, is not surprising. In the hierarchy of newsworthiness, such a diving drama is probably accurately accorded a fairly high position. Not only does it contain an important element of suspense, the race against time to rescue those trapped below, but also it strikes a responsive chord in a public imagination that is readily fascinated by the apparent derring-do of an exceptionally high-risk occupation. Moreover, by providing a cue for discussion of the dispro-portionately high risks encountered by those employed as divers in the North Sea, such an event offers reporters a chance to link the drama of the specific incident to wider issues. Every one of the accounts referred to above, for example, contained a brief analysis – usually different from that offered by competitors – of the number of divers reckoned to have been lost in the course of offshore oil operations during recent years. Indeed, in some cases such broader calculations not only promp-ted reflection upon the statistical record of offshore safety in general, but also resort to the familiar rhetoric of disaster. 'North Sea oil toll nears 100' was the caption which introduced the *Observer's* discus-sion of this wider theme some four days after the accident.[3] The use of such language, evocative of the way in which the mounting toll of an earthquake or similar disaster might be presented, provides an exam-ple of the powerful images that are available to connect the hazards confronted by offshore workers with potent cultural themes.

Pre-eminent among such images is that of an industry facing all the risks associated with operating at the outer limits of existing techno-logy and in physical conditions which are, at the very least, consistently more difficult than anywhere else in the world.[4] As Guy Arnold remarks, 'North Sea oil has created its own language: the leading jargon phrase is "on the frontiers of technology" and everyone loves the image this creates.'[5] Similarly, even the industry's own project engineers are said to have been left breathless by the pace of technolog-ical developments which have been spurred along not just for the sake

Images of Danger

of change in itself but by 'the peculiar circumstances that engineers
have found themselves dealing with in the North Sea'.[6] Nor is there any
dearth of reports which link the challenge of the offshore enterprise in
these respects to its implications for safety:

> The North Sea has presented man with one of his biggest challenges
> this century . . . a wealth of energy lying 12,000 feet below turbulent
> 600-feet deep seas. An area which can produce winds of up to 160 mph
> and waves of 100 feet – as tall as an eight-or nine-storey building. It is
> here that the offshore industry has constructed the necessary
> platform giants to extract oil and gas and send it ashore. . . . In such a
> hostile area, safety and the environment provide the starting point
> for all decision-making. . . .[7]

Whatever the priority allocated to safety in the planning and execution of offshore operations, the frontier image is one that readily
reconciles readers to the inevitability of accidents. People are killed at
inhospitable frontiers. Thus a further image which is often projected in
discussions of offshore safety is that of necessary sacrifices for the
common weal. While we may dismiss as mere, if possibly inappropriate, humour the view of one Treasury man who reportedly
explained to an early collaborator on this project that 'with the
economy in the state that it was, the people who were dying there
were dying for the greater good',[8] the suggestion of necessary sacrifice
is often quite explicit and quite serious. As we have already seen, for
example, when Mr Benn addressed the subject in 1977, he not only
chose terms more redolent perhaps of the Cenotaph than of the North
Sea, but also alluded to the tragic 'price' so often extracted from
pioneers.[9] More recently, the *Guardian* deployed the metaphor of
cost–benefit analysis to provide a stark caption for a discussion of 'The
human price of Britain's oil billions'.[10] Sometimes even the sense of
metaphor disappears, as when the Head of Department of Energy's
Petroleum Engineering Division explained that the offshore oil industry would show up very favourably if comparisons with other
industries like mining were based on the amount of energy produced
per fatal or other serious accident, or per inspector. For the media,
however, the metaphor usually suffices, as can be seen from the following extract from an editorial which combined the idea of a 'cost' with
several of the other images mentioned above:

> North Sea oil and gas will be worth a staggering £7,200 million this
> year – and we owe an enormous debt to all who have helped to bring

this about. The costs have been high and grievously so in regard to the loss of life which has been incurred. Only yesterday we had another grim reminder of the human toll involved in this vast and crucial operation, in often severely testing conditions which demand taking technology to its outer limits.[11]

This portrayal of death and injury as a 'price' or 'cost' which should be entered in the balance sheet is not something that is often so directly articulated in connection with employment onshore. With reference to North Sea oil, however, it is a recurrent theme and one which, moreover, combines readily with the 'reckless and ruthless' image conjured up by the use of terms like 'oil rush' and 'oil strike'. On the one hand, such terminology facilitates the emergence of a stereotype of the offshore worker as someone who, like his precursors in other strikes and rushes, recklessly embraces immense hardships and incalculable risks in the pursuit of quick rewards, a sort of personification of the broader calculus whereby the nation's economic ills may be at least partially cured by the risky North Sea venture. As one senior departmental official put it,

> We are talking about the type of lads who chase money, the kind of fellahs who are prepared to go up to the north of Scotland and work on hydroelectric stations for maybe four months without getting home or anything like that, because they have got a target of £2000 or £3000 in mind – a certain type of labour which hasn't shown itself in the past to be very careful about how to do things.

Inevitably too, this 'cowboy' image spills over into a familiar representation of the offshore worker's enthusiastic preparations for a prolonged period of complete deprivation of some customary pleasures. At Dyce Airport, as one reporter with a particular flair for the colourful put it, 'The rig crews roam in and around the bar drinking like human blotting paper while their helicopters crouch outside waiting their next cursing load.'[12] A car sticker proclaiming to the world in neatly ambiguous terms that one particular company's divers 'do it deeper' adds another essential element of *machismo* to the image enjoyed by at least one group of offshore workers.

On the other hand, of course, the idea of death and injury as a cost or price of exploiting offshore energy reserves also merges readily with the image engraved upon the obverse of the oil-rush coin – that of an industry which pays but scant attention to the human cost of its financial returns. In 'The Cheviot, the Stag and the Black, Black Oil', 'a ceilidh play with scenes, songs and music of Highland history from the Clearances to the oil strike', for example, the American oil man calls the square dance moves:

> Pipe that oil in from the sea,
> Pipe those profits – home to me.
> I'll bring work that's hard and good –
> A little oil costs a lot of blood.[13]

By the end of 1978, 'piping the profits home' was already playing its part in creating the improbable spectacle of a country, by then around 80 per cent self-sufficient in oil, running up a £2.5 billion deficit on its oil balance of payments in a single year.[14] Similarly, back where home is for most of the oil majors, market solutions to the so-called energy crisis were shortly thereafter said to be recycling income towards the industry on a scale so vast as to create 'a problem unparalleled in the history of literature' and a situation where 'the companies are no longer able to buy talent capable of justifying with even minimal plausibility the increases in their profits'.[15] More generally, the impact of the sheer scale of the world-wide oil transaction upon the industry's image has been succinctly expressed by Doran, albeit by way of outlining a myth he subsequently purports to demolish:

> Naturally, the size of the oil companies makes them highly visible despite conscious low-profile strategies. Incomes of the largest companies dwarf the national incomes of many of the countries in which they operate. The average incomes of oil company employees exceed by a large margin the average incomes of citizens in the host countries. Ineluctably, the image of the oil company becomes that of wealth, exclusiveness, secrecy and dominance.[16]

Back in the North Sea, meanwhile, the 'lot of blood' being expended in the search for oil was not passing entirely unnoticed. In 1975, for example, the Scottish Council of the British Medical Association (BMA) investigated the medical implications of oil-related industry, and came to some stark conclusions about the offshore sector. The fatality rate per 1000 rig workers in the North Sea, it calculated, was approximately twice that for deep-sea trawlers in near and middle waters, ten times the rate for coal mining, more than ten times that for the construction industry and fifty times the rate for factory employees.[17] These conclusions were subsequently to be reported in journalistic accounts, in legal textbooks on offshore employment and

even in reports by the US Occupational Safety and Health
Administration.[18] Moreover, while such an august body as the BMA
was obviously content to state its conclusions without derogatory
embellishment, other commentators were not slow to see a connection
with the industry's reputation for ruthlessness. 'Fall off that derrick,
buster, and you're fired before you hit the deck' was the quotation used
to preface one chapter in a legal work on the subject.[19] 'In my [book],
reports another putative author, '[it is] if you fall off an oil rig they
don't throw you a lifebelt, they throw you your books.' In similar vein,
while Mervyn Jones's and Fay Godwin's beautifully illustrated book
The Oil Rush did much to dispel the myth of the 'cowboy' worker, it
did little for the image of the ruthless company. Indeed, with its
chronicle of offshore workers greeting official casualty statistics with
bitter laughter, of injured employees remaining uncompensated, not to
mention unemployed, and of drilling companies even declining to
suspend work when someone was killed, this impressionistic account
of North Sea operations amounted to an indictment of the industry's
callousness.[20] It was left to the *Scotsman* to express what it called this
'shabby side of the oil boom' in the more measured terms of economic
metaphor:

> What price North Sea oil? It can be measured as the reserve cost,
> which is the effort the oil companies need to put in to get it out; or as
> economic rent, the revenues which will accrue to public funds as a
> result of the activity; and it can be measured in terms of oil company
> profits. One vitally important cost factor, however, is often neglected
> in all the studies and assessments of the world's newest major oil
> province – the human element.[21]

Encounters of the Dangerous Kind

The pattern of offshore accidents revealed in an earlier chapter would
seem, at first sight, to lend considerable substance to the image of
North Sea dangers as an inevitable by-product of a very special
industry facing special technological problems within an especially
hostile environment. The very special risk of death among divers, for
example, might seem to reflect what is only to be expected when such
operations are undertaken at such great depths and in such low
temperatures. Drilling, with its complex requirements of draw-works,
derricks, rotary power and high-pressure injection systems, not to
mention the need for split-second timing in making connections and

manoeuvring lengthy sections of heavy pipe into position, seems to bring together a formidable series of hazards and, in many ways, encapsulates the industry's rugged image. Similarly, the record for crane and vessel operations fits well with the image of an enterprise which has constantly to contend with a hostile environment that renders activities like unloading, mooring and anchor handling extremely difficult. Indeed, this picture of danger emanating from a running battle with the elements becomes even sharper when it is realized that no fewer than fourteen out of the thirty-eight non-diving deaths reported between 1974 and 1978 involved the deceased person, for one reason or another, falling into the sea.[22] And on top of all this, of course, the accident pattern which emerged could easily be represented as little more than a reflection of the risks inevitably associated with operating at the outer limits of knowledge – whether in the engineering impertinence of erecting massive structures in such a setting, in pushing diving to its physiological and technological limits or in drilling for, and then controlling, the highly volatile substances lying so far beneath the surface.

The official record of accidents which have taken place in the course of North Sea oil and gas operations could all too easily be taken as a reflection of what is only to be expected when an industry undertakes inherently hazardous operations at the very frontiers of technology and in the face of appalling climatic conditions. Against such a backcloth, it might seem that the routine application of laws relating to safety would have little useful role to play in limiting the risks involved in offshore employment.[23] But just how special is the North Sea in these respects? More precisely, to what extent is its catalogue of death and injury the result of the unique exigencies facing the offshore oil industry? To answer this question, we must go behind the bare story told by statistical patterns and examine offshore accidents in somewhat greater detail. By so doing, we will better be able to assess the extent to which some of the images discussed earlier in this chapter are true likenesses.

In the course of this research a number of people were encountered who were at pains to stress that, despite the aura of high drama surrounding offshore operations, many, if not most, of the accidents which occur are the result of much more mundane factors, well-known to anyone who is familiar with the world of industrial safety onshore. Thus, for example, an Aberdeen surgeon who deals with many North Sea casualities emphasized that the injuries passing through his care were very much what might be expected in any major industrial context onshore and largely the result of similar causative factors – 'dropping things, having things dropped on them, falling from a height and so on'. Similarly, I was told that an internal company survey of accidents in the Forties field had revealed 50 per cent to be the result of 'falling, slipping or tripping', not exactly the stuff of which the North Sea's image is made. The point was put most cogently by a Health and Safety Executive inspector:

> I am quite convinced that offshore there are just as many accidents of people tripping over things, cutting their hands, sticking their hands in V-belts, getting tangled up in drilling machines – these sort of things – burning themselves with welding and gassing themselves, all the things they do in industry. . . . Look, the vast majority of accidents in industry are caused by relatively minor incidents. The number of people who are actually involved in major incidents are very small and you could make a far bigger impression in getting the small things right.

Examination of unpublished accident analyses carried out annually by the Department of Energy lends some support to this more mundane version of events. Between 1975 and 1978, the period for which these data have been made available, around 30 per cent of the 229 fatal and serious accidents which were reported resulted from falls, while the broad category of crushing accounted for a further 25 per cent. Some corroboration of this pattern in the causation of oil industry accidents elsewhere than in the North Sea is provided by a report from the International Association of Drilling Contractors, Houston, Texas,[24] while a recent Occupational Safety and Health Administration survey of thirty fatalities occurring on drilling rigs in the United States found that no fewer than twenty-three of them were the result of falls from elevations.[25] Back with the North Sea, even a cursory reading of the slightly more detailed quarterly accident summaries made available by the Department of Energy leaves a strong impression that, falls apart, the vast majority of offshore injuries stem from relatively conventional causes, even if the scantiness of the information contained in these brief accounts precludes further quantification.[26]

As the second of the two American reports referred to above points out, establishing the aetiology of accidents requires more in-depth investigation than can be accomplished from classification by codes and so on. More specifically, it stresses that narrative and other information contained in investigation files are an 'underutilized source of accident data', even though they provide 'more detailed descriptions of how occupational fatalities occur than do any other data source currently available'.[27] The experience gained in the course of the present study would tend to support this view as far as oil-related accidents in the North Sea are concerned,[28] fatal-accident files maintained and made available by the Scottish legal authorities proving to be an invaluable source of data.[29] All such files which were available in mid-1978 were examined,[30] the total number of fatalities covered coming to forty-one. The series included thirty-two out of the thirty-nine accidental deaths listed in the Department of Energy's quarterly summaries and annual statistics for 1974 to 1976, as well as six fatalities which occurred outside this period and three others which, for one reason or another, fell outside that Department's jurisdiction. The total number of accidents, as opposed to deaths, which was covered came to thirty-nine.

In what follows, the story of how these fatalities came to happen will be largely allowed to speak for itself. (Readers who are not familiar with some of the terminology are referred to the Glossary, pp. 310–311, where some of the more important terms are briefly explained.)[31] Each of the reputedly most dangerous categories of offshore operations will be dealt with in turn, and for all but one class of accident – diving, in connection with which there are relatively few serious accidents which stop short of the fatal – fatality information will be augmented by a series of brief descriptions culled from the Department of Energy's quarterly summaries of serious casualties. In order to provide approximate uniformity in the number of cases of the latter type to be presented, *all* of those reported during either the entirety or an appropriate part of 1978 will be included. The result, it is hoped, will provide convincing documentation of the thesis that most offshore injuries result from comparatively conventional causes. While there is no intention or desire to deny the technological sophistication of the North Sea oil industry or the difficult operational conditions which it has to face, these factors, I contend, do not bear primary responsibility for the safety record of the British North Sea to date.

Construction

Although construction accidents did not top the statistical league table of offshore risks discussed above, this phase of the North Sea exercise is often held to be fraught with special danger.[32] Even after a platform has been placed in position, a great deal of major construction work still remains to be done, and indeed the prolongation of this kind of activity beyond the original estimates of most licencees has been a cause of concern to the Department of Energy.[33] Not only does this involve accommodation problems, because construction workers have to remain on board longer than predicted, but it also involves questions of exacerbated risk arising from the simultaneous performance of difficult operations under more cramped working conditions.[34] And, of course, the task itself – dangerous enough onshore – has to be carried out in the inhospitable environs of the North Sea. As one doctor, clearly impressed by a visit to a platform under construction, pointed out, 'It was extraordinary there were so few accidents relative to this situation where, I mean, you could hardly walk along, you could hardly find a clear way to walk on. . . .' A Procurator Fiscal put it succinctly when he wrote, 'It is notorious that platforms in the course of construction are fraught with danger for the unwary.'

Evidence from the thirty-nine fatal accident files lent some substance to the belief that offshore construction work does indeed involve rather special hazards, inasmuch as the only two completely unexplained deaths in the entire series involved construction workers who simply disappeared, presumably lost overboard (Cases 29 and 36). But the other two accidents to claim the lives of such employees were somewhat less out of the ordinary. In one case, the deceased person fell while descending from a temporary accommodation ladder, which, in the words of an investigating police officer, 'was totally lacking in any form of safety device'. According to the Department of Energy inspector who went out, it was 'an accident waiting to happen' (Case 21). In the other incident, a gin-pole was being dismantled after the construction of a maintenance crane on the flare-stack floor, some 200 feet above the main deck, when it fell across the floor and fractured the spine of a site engineer. Without specific instructions or any supervision, two employees had proceeded with a task described in the accident report as 'dangerous if less than four men are engaged', and had replaced a $^3/_8$-inch wire lashing, capable of carrying a weight of almost 10 hundredweight, with a $^1/_4$-inch one which, although probably able to take 5 hundredweight, proved no match for the 7.5 hundredweight actually involved (Case 32). While the web of misunderstandings, and indeed of good intentions, lying behind this incident cannot be elaborated here, the case serves to underline the fact that the special exigencies of the North Sea do not always have to be held accountable for construction accidents. So too do the following summaries of the six serious construction accidents which were reported during 1978:

1. Fitter – Burns to hands, face and neck.
   A parts container that is used as a workshop on starboard side . . . has a propane gas heater. The victim struck a match to light his pipe and there was an explosion.

2. Two welders – Burns to nose and cheeks, and to face, scalp, wrists and hands, respectively.
   Whilst welding and grinding there appeared to be a flash fire which ignited tarpaulin and scaffolding battens. . . .

3. Welder – Two fractured ribs and abrasions to legs.
   The casualty was carrying a welding extension cable when he fell 6 feet between a gap in modules. . . . The cause of fall was a combination of lack of personal care and conflicting lighting (bright spotlights which almost blinded one, and in other places, dark shadows).

4. Rigger – Broken leg.
   While lifting a pipe into position using an air-powered winch, the winch came free from its four L-shaped retaining lugs which had been welded to the deck plating. It struck the victim as it was pulled along by the weight of the load.

5. Scaffolder – Hypothermia and multiple lacerations to head.
   Man descended on to suspended scaffolding to remove fittings. On standing on scaffold platform, the outer board broke.

6. Scaffolder – Fractured pelvis, neck and left radius.
   Erecting scaffolding . . .when scaffolding gave way, causing a fall of approximately 19 feet 9 inches to the deck.

Drilling

Drilling is the central activity upon which all else depends in the search for, and extraction of, North Sea oil and gas. As we saw at an earlier point, it combines a considerable number of hazards and has thus far contributed its full share to the record of offshore accidents reported to the Department of Energy. Indeed, such are the risks of being injured in this context, that a high accident rate is sometimes accepted as an inescapable fact of life as far as drilling activities are concerned:

Very often when we are discussing statistics, we'll say to our platform manager, 'Well, your figures are high, higher than they ought to be.' And he'll say, 'Oh well, we've got drilling going on.' And I say, 'Well, but you're not really telling me that you are willing to accept a drilling accident but not another type.' (Safety Manager)

Once again, however, there is evidence to support an approach which emphasizes the 'ordinariness' of drilling accidents. According to the safety manager quoted above, for example, 'The drilling fraternity do have a higher frequency rate, but again the type of accidents they have are the same type of slipping and being crushed and banging into things.' Similarly, examination of the material made available highlights just how frequently such accidents seem to result from fairly conventional factors such as poor communication, failure of equipment, poor working practices and lack of safety precautions. Indeed, of the eight drilling fatalities in the series, only two did not fall fairly clearly into one of the above categories – one driller who unaccountably stepped forward into the drill hole, having himself removed the temporary cover (Case 27a), and a floorman who was killed by a 45-lb piece of wooden bumper which fell on him during an attempt to repair previous (and unreported) damage to the crown block at the top of the derrick (Case 18).

Three of the six remaining drilling deaths on which there is detailed information resulted from straightforward errors and failure in communication. One derrickman fell 85 feet from his monkey-board when a collar stand was lowered into the drill hole before he had detached the tugger line used for manoeuvring pipes and collars into position. He was not wearing a safety harness even though the firm had been warned about slack safety discipline on the drill derrick only two months previously (Case 7). Another fell 40 feet from the stabbing platform after he had attached the spider elevator to a column of casing a fraction of a second before the power tongs had applied the final torque needed for a satisfactory connection (Case 2). A serious breakdown in communication caused the death of a driller who entered the housing of the draw works, a large horizontal winch on the drill floor, without informing the operator who subsequently started the winch, thereby crushing him (Case 12).

The other three drilling fatalities in the series resulted from constellations of factors which will be no less familiar to those acquainted with how easily accidents can happen in any industrial setting. There is nothing uniquely 'offshore' about a driller falling from a 30-foot vertical access with several steps, which were subsequently described by one witness as 'so badly twisted that it is difficult to get a foothold between the rungs' (Case 30). Nor is it exactly the frontiers of technology or a hostile environment that are being confronted when a butterfly valve rated at 175 pounds per square inch (psi) maximum safe working pressure is attached to a chicksan line with a 10,000 psi rating while, in order to speed up operations, fluid is jetted through by a positive displacement mud pump working at around 1750 psi (Case 31). Similar shortcuts to save time, even sometimes at the expense of safety, are by no means unknown in industry ashore. Such a shortcut was put forward as the probable explanation for one final drilling accident which killed the driller himself. Drilling line had failed to rewind evenly on to the draw works and, in order to rectify this potentially damaging situation, the line was being pulled out in a kind of 'tug-of-war' exercise. For this to be accomplished, it was necessary to lower the drilling block to the drill floor, thereby taking tension off the line. But an order to secure the block against unanticipated movement by means of a safety line was countermanded, and it was left supported by its hook which was resting partly on the floor and partly against some protruding pipe (see figure 1). When the hook slipped, the block and compensator, weighing 85,000 lbs, fell 2–3 feet, a distance increased tenfold by the pulley system, and the line immediately tightened, catapulting the deceased head-first into the draw works (Case 8).

Dramatic as the tragic consequences of such accidents certainly are, this more detailed examination of their aetiology suggests that we should be cautious about accepting either the inevitability of a high fatality rate in drilling or, indeed, the idea that we are dealing here with some mystique, impenetrable to all but those who have worked in, or closely with, the oil industry. While human error will never be eradicated, the majority of the above deaths could almost certainly have been avoided if fairly obvious precautions had been taken. Equally, to make a point that will be dealt with at greater length in a subsequent chapter,[35] there was nothing involved in any of these accidents that would have fallen outside the competence and comprehension of a factory inspector. The same can be said of other serious drilling accidents, as is evident from the following summaries of those reported during the last three months of 1978:

1. Floorman – Dislocation of left shoulder.
     The victim was cleaning out the slush pit and waiting for a ladder to be passed to him in order to climb out. He had one

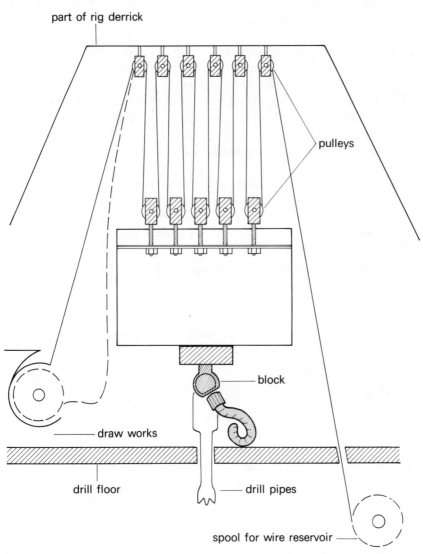

part of rig derrick

pulleys

block

draw works

drill floor

drill pipes

spool for wire reservoir

Dashed line = slack wire; solid line = taut wire

Figure 1
Pulley system of drilling block

foot on the agitator when it was accidently switched on, throwing him against the tank wall. Safety pins had not been used to prevent accidental activation.

2. Derrickman – Fractured head of radius and scaphoid, right arm.
   The derrickman stood on top of the blow-out preventer (BOP) to change the position of the slings (during removal operations). The drill block was lowered and BOP rotated through a horizontal axis, hitting him. He grabbed an overhead chain but thought the BOP was going to fall and took evasive action, falling a distance of approximately 18 feet.

3. Roustabout – Soft tissue injury to lower back area.
   Removing snubbing hook from Kelly bushing, victim slipped and fell on his back striking lower back on mousehole cover. The cover guides caused stab wounds in lower lumbar area.

4. Floorman – Amputation of fourth and fifth toe of right foot.
   Chicksan current pipe sheared off and fell on victim's foot.

5. Floorman – Fractured leg.
   Laying down pipe – struck by a ruptured connecting link from a chain and hoist line.

Cranes

Cranes obviously play a vital role in North Sea operations since it is upon them that any installation has to depend for the transfer of equipment and supplies from vessels, for moving heavy materials from one location or level to another, and sometimes even for the transfer of personnel by basket. Such operations do, indeed, present special problems in the offshore setting. Lifting heavy loads is a difficult enough task in its own right, without added complications such as a heaving ship's deck, which can impose shock loadings in a severe swell. If the load should snag on the vessel's superstructure or become immersed in water, the problem is compounded. Moreover, salt water or spray can have a corrosive effect on both wire ropes and exposed parts of crane machinery, thereby increasing the likelihood of equipment failure in an emergency. This possibility is accentuated by the fact that most offshore cranes are pedestal-mounted and therefore particularly vulnerable to boom collapse or even to the cab's being torn from its base if the crane is overloaded. The special exigencies of the North Sea then add one final twist of fate, since there is a strong possibility that in the event of the latter occurring, the cab and the operator will go overboard.[36]

Against this background, it is tempting to view crane operations and their associated dangers as yet another uniquely offshore problem, which can be understood only by the fully initiated and solved only by further technical advances (for example, by the introduction of dampers to reduce shock loading). The former was clearly implied, for example, by one safety officer, who insisted that 'whoever dreamed up the idea of requiring overwind trips on offshore cranes had obviously never been near an oil-rig', since the activation of such a device precludes the operator from 'paying out' quickly in order to rectify a sudden overload. Similar conflicts between those 'in the know' and those who construct regulations have arisen over a Norwegian ban on personnel baskets except by special permission of the Norwegian Petroleum Directorate.[37] In the British sector, offshore custom has in the past collided with established onshore practice over the use of cranes with free-fall facilities for carrying personnel – to the point where, as one Health and Safety Executive inspector explained, 'They either stop personnel carrying offshore, or they go against our guidelines which everybody uses on land.'

At first sight, the five files pertaining to crane-related fatalities which turned up in the series would seem to support a view of such accidents as the inevitable, if undesirable, consequences of operating in a setting such as the North Sea. Two deaths resulted from cranes falling overboard during unloading operations, one from a crane collapsing on to the deck while trying to lift sub-surface piping, and one from a roustabout being picked up and dropped overboard while attempting to connect a crane hook to a mooring line. With the exception of the latter case (Case 14), however, closer scrutiny of these and of the one remaining death associated with cranes showed that while the very special characteristics and problems of offshore operations cannot be completely exonerated, there were also other and more mundane factors at work.

In two of the instances in which this was most immediately evident, operator error seems to have played a crucial role. One crane driver dropped an empty bin on a roustabout when he engaged lowering gear and let off the handbrake without engaging the clutch or having his foot on the footbrake. Although an experienced operator, he was not familiar with the type of crane in use and had been receiving instruction from other operators during the four days preceding the accident (Case 4). In the case of the crane which collapsed on to the deck, thereby killing its driver, operators on the platform in question were not aware, apparently, of an override button which would bypass the overload cut-out if this should be necessary. There was also some suggestion that an experienced operator would have been more aware of the need to observe the load indicator when performing an unusual task such as that being undertaken at the time of the accident (Case 35).

The two cranes which fell overboard provide good examples of how the admittedly dangerous qualities of the North Sea can be rendered fatal by other circumstances which are by no means exclusively the prerogative of that operational location. In the first case, a 5-ton load was being lifted from a supply boat when, according to one witness, the brake failed and the load fell into the sea. When it was raised from the water, the same thing happened again. At the third attempt, efforts to stop a further repetition led to its descent being stopped very abruptly, whereupon the crane slowly toppled over the side, hitting the sea at the point where the operator had just landed in a vain attempt to get clear. It was claimed that similar brake failures had occurred on four previous occasions, the most recent being earlier on the day of the accident. Although confirmation (or refutation) of any deficiencies in this respect could be obtained only if and when the crane is recovered from the sea bed, subsequent metallurgical analysis of what had been left behind showed that the welding of a strengthening plate had not been satisfactory, that flame cutting on one edge of a bracket had removed this weld, and that, as a consequence, the plate had been prevented from carrying its share of the load and had failed completely in the overload situation which arose. Damage on the rollers was said by one expert to have occurred prior to the accident and, although having no bearing on the actual incident itself, was thought to indicate lack of careful in-service examination that might have revealed the other defect, which did have such a bearing (Case 19).

Reports on the remaining fatality of this kind revealed a picture amounting almost to complete chaos. Again, supplies were being unloaded, a 15-ton drum of wire having just been lifted clear of the boat's deck at the operative time. The vessel surged forward on a wave and caught the drum under the projecting lip of its high tailgate. As the supply boat fell into the subsequent trough, the crane started to topple. While the operator had time to jump clear of its ultimate landing point in the sea, he was thwarted in his attempts to reach knotted ropes thrown into the water from the rig's control area because the supply boat steamed full-speed ahead away from the installation, snapping its retaining warps and removing the vital assistance which its propellor wash had hitherto been rendering in this respect. At this point, a mechanic on the rig started to lower its only serviceable lifeboat, but was told to desist. Shortly afterwards, however, the installation manager appeared and ordered him to proceed. The lifeboat was launched, but the after-fall line had taken a bight round the rudder, damaging it to such an extent that although the boat could go ahead and astern, it could not be steered. Apart from the crane operator, who by this time was beginning to 'lose interest in the proceedings', there were now four other men effectively adrift close to the rig. The stand-by boat eventually appeared on the scene and, after some confusion over who should throw a line to whom, took the lifeboat in tow, pulling it to the supply vessel, which then took over the task of towing it back to the rig. The lifeboat then collided with the supply boat, its hull being split from deck centre to keel, and the supply boat's anchor crashed through its deck. Scramble nets were finally put over the vessel's side, and the lifeboat crew successfully abandoned their craft. No trace was ever found of the crane operator.

With suitable embellishment of the stark details recounted above, an incident like this could almost have a place in some anthology of sea adventures. But the background which emerged from subsequent investigations produced a more sombre and familiar tale all round. The crane was fitted with only a single front hook roller, whereas two would, at the very least, have slowed its rate of collapse. A sister crane on the same rig had suffered a similar accident, though not as catastrophic, some time previously and had subsequently been modified in this way. Indeed, a crane engineer expressed surprise at finding an unmodified crane of this kind still operating in the North Sea, since the manufacturers had been replacing single with double hook rollers for the past seven years. Moreover, the drum of wire had been lying on its side and therefore in a position from which it 'should not have been lifted at all'. Had it been on edge, and a bar attached to slings placed through its centre, the flanges – designed to catch on protruding edges – would have rolled over or up the edge on which they caught. Why too, it was asked, was the safety boat not in closer attendance, and why was *it* not capable of launching a rescue craft at once? Why, finally, was the crane operator not wearing a life jacket, and why did he ignore recommended practice by discarding some clothing and attempting to swim? The investigation's conclusion that the whole affair was 'all in all not a very good picture' would seem euphemistic (Case 17).

Once again, then, the temptation to put a specific form of offshore danger down to unique operating conditions and technical problems has to be resisted. Such there certainly are; but, as we have seen, they by no means tell the whole story. Indeed, the more detailed investigations outlined above would seeem to lend broad support to the view, attributed to the senior mechanical inspector at the Department of Energy, that 'a large number of offshore crane accidents are due to negligence and carelessness rather than to any inherent weakness in the cranes themselves.'[38] While we do not have anything like the amount of detailed information that would be needed before any similar conclusion could be reached about non-fatal crane accidents, the following summaries of the four serious ones which were reported during 1978 would, with one possible exception, seem to be describing fairly routine lifting-gear incidents:

1. Rigger – Lacerations to arm and fractured ribs.

    Lifting a container with a fork-lift truck. The container caught a projection of metal sheets and fell, striking the victim and trapping him against the handrail.

2. Roustabout – Fractured right scapula and bilateral perforation of eardrums.

    When transferring a tank to the forward starboard hose guides by crane and trying to turn the tank into correct position prior to landing, the injured person was trapped between bulkhead and tank. The load swung due to the motion of barge.

3. Roustabout – Traumatic amputation of middle and index fingers of right hand.

    The crane was being used to remove equipment from the roof of the Quarters annexe. The hook of the crane caught on the handrail surrounding the roof of the annexe. The injured man attempted to free the hook while the crane was still in motion trapping his fingers between the hook and the handrail.

4. Painter – Complicated fracture of right leg and abrasions to left leg.

    Moving sheets of steel utilizing crane and plate clamps. Injured person was steadying sheet when clamp grab slipped. He stepped back but tripped over other sheet, and the sheet he had been holding fell across his legs.

Vessels

In chapter 2, we saw that accidents involving personnel on vessels have played no small part in the generation of offshore casualties. Although it was not possible to produce incidence rates for deaths associated with the specifically oil-related operations of vessels in the North Sea, this is not suprising. Manoeuvring close to installations in heavy seas, unloading weighty pipes and containers on to platform decks a long way above while maintaining position against tide and current, performing the vital task of raising and relocating massive installation anchors – all of these are activities whose dangers scarcely need underlining. And on top of all this, of course, there is always a constant risk from the sea itself.

Although information about vessel accidents is, for various reasons, scantier than for any other category of North Sea incident, it would seem that this class of accident comes closest to fitting the image of danger emanating from the special circumstances of offshore operations. No fewer than four of the seven fatalities on vessels appearing in the series of available files were attributed directly to sea conditions. One man was washed overboard by a 'huge wave' and another by a wave of unspecified dimensions; two seamen were crushed by or against deck cargo as a result of what were referred to respectively as 'a large sea' and 'a freak double wave such as has been described in a number of maritime accidents' (Cases 15, 16, 20 and 27). Two of the remaining deaths took place in the course of operations to raise installation anchors – one when the winch cable crushed the ship's mate against a marker buoy stowed on the deck, and one when a 'pelican hook', designed to take strain off the winch, fractured and caused the anchor cable to whiplash (Cases 1 and 6). The final fatality resulted from a 95-foot walkway between a derrick barge and an installation under construction rolling forward and crushing a rigger against the rails (Case 23).

A similar pattern is evident from the serious accidents reported during 1978:

1.   Derrick foreman – Compound fracture of left tibia.
     Working on deck of [vessel]. Winch wire broke and lashed back, breaking the victim's leg.

2.   Two seamen – Fractured pelvis; internal abdominal injuries.
     While attending [sic] the slings of a container weighing 6

tons . . . to the hook of a crane . . . the two injured men had placed the lifting ring over hook which was hanging down the side of the container. The [vessel] fell into a trough in the waves, thus causing the crane to simulate hoisting. The two men were trapped betwen the container side and the sling/hoist rope of the crane.

3.   Two seamen – Fractured skull, head wound.
     While working on the moorings of the rig, two crew members of the vessel were struck over the head by an anchor wire.

4.   Seaman – Broken left leg.
     When taking tugger wire from main wire connected to the ring's anchor pennant, the main wire became taut, throwing the victim across the deck.

5.   Able seaman – Broken collar bone and three ribs.
     The victim was working on deck of supply boat preparing to offload cargo. . . . A wave broke over the stern of the ship, causing container to shift position. The victim tried to get behind the crash barrier, and in so doing was crushed between the container and the crash barrier.

Although the above pattern seems, as already suggested, to be broadly in line with an interpretation of such shipboard accidents as a consequence of peculiarly difficult operations having to be conducted in a frequently ferocious setting, several qualifications need to be borne in mind, even if paucity of data precludes elaboration at length. For one thing, several of the fatalities referred to earlier do seem to have involved other factors which could possibly have been crucial in failing to prevent the accidents in question. The fractured 'pelican hook', for example, had been 'bought off the shelf', and whereas the purchasers had specified a 50-ton capacity – on the assumption that this meant a safe working load of 50 tons – the manufacturers' standard practice was to mark such hooks with their proof load. In this particular instance, the hook had been tested to a *proof* load of 50 tons, but its safe working load had been certified as only 25 tons. Whether or not this discrepancy actually caused the accident, however, could not be satisfactorily determined. A similar aura of inconclusiveness surrounded investigations into two of the other deaths. According to the Sherriff's formal determination in the case of one of the seamen washed overboard, 'Part of the area behind the sea wall was not fenced by rail and part was, but it is not established from which part the deceased was swept overboard.' Similarly, in the incident involving the moving walkway, there was an unresolved conflict of evidence as to whether clear, or indeed any, instructions to stand aside had been given before an attempt was made to hoist the gangway into position.

Reservations must also be entertained with regard to the frequency with which large, huge or freak waves are apparently implicated in these kinds of accident. As one sceptical Health and Safety Executive inspector put it, by way of hyperbole if not sarcasm, 'There seems to be a freak wave every five minutes in the North Sea.' It is also worth recording in this connection that one subsequent fatality was encountered in which the accident had been attributed to just such a wave until two witnesses turned up at a Fatal Accident Inquiry and gave evidence to the effect that, for some time, part of the guardrail had been missing from the point at which the drowned man went overboard. If possible deficiencies such as these and the ones referred to above do play a significant part in vessel accidents associated with North Sea operations, the assessment of their 'special' causative characteristics would have to be revised. Indeed, the issue would become inseparable from the broader concern which has recently been voiced about safety on merchant ships in general.[39]

Diving

Diving has been left to the end because it is usually considered to be, by any standard, special. Here, more than with any other North Sea activity, we would expect to find a pattern of accidents reflecting occupational confrontation with the very frontiers of technology, physiological endurance and environmental hostility. As reference to the Glossary, where some of the techniques and terms utilized in diving are outlined, will show, moreover, the frontier image of this dangerous occupation is based on no idle boast. But does it entirely account for diving's appalling safety record in the North Sea? To answer this question, we will examine the circumstances surrounding nine diving fatalities which it has been possible to investigate in some detail. Since serious, as opposed to fatal, diving accidents are comparatively rare, at least according to the official statistics, I will depart from the previous practice of summarizing a sequence of these at the end of the section.

In early 1975, two divers carried out an operational dive to a depth of around 380 feet for the purpose of clearing rope and wire cable that had fouled a blow-out preventor stack. To accomplish this, they descended in a bell which was pressurized to the appropriate depth with the necessary mix of around 5 per cent oxygen and 95 per cent helium. Because of the well-known problems of hypothermia (cold) associated with North Sea diving, steam heating in the deck decompression chamber to which they would return was turned on. After the dive was completed, the bell returned to the surface, and some difficulty was encountered in establishing a good seal between it and the 'transfer-under-pressure chamber' through which the two men would pass on their way to the main chamber. At a second attempt a satisfactory connection was made, but as the divers were in the process of changing the bell controls, a slight loss of pressure was noted. Although this was corrected by two short bursts of 100 per cent helium, a further slight loss of pressure suggested to the diving supervisor that the original problem with the connection was recurring. Accordingly, the divers were ordered into the transfer chamber and told to seal off the bell. The latter they were unable to do, and, since a small pressure loss was still registering, they were instructed to proceed to the main complex and to shut the door.

On entering the main decompression chamber, the two men complained of the heat; the heating, which had earlier been registered at a high but not unacceptable level of around 110°Fahrenheit, was switched off. But a loss in pressure was still being recorded on the master gauge, and more helium was therefore introduced to the main chamber. This still did not improve the reading, so the supervisor now concluded that he was facing a major emergency. At this point also, he suddenly realized that he was, in fact, monitoring not the decompression chamber but the bell, and when the necessary gauge alterations were carried out, it became evident that the series of helium injections into the former had re-pressurized it to a depth of 650 feet. This in itself was not particularly dangerous, but pressurization increases temperature, while the temperature conductivity of helium feeds heat to the human body at a high rate. The humidity of the chamber's atmosphere had also increased to near saturation point. Both divers showed increasing signs of acute distress, possibly to the point of panic, but there was no means of accelerating the pace of temperature reduction, even though the real nature of the emergency was now clear. Shortly afterwards, the victims were seen to collapse on the chamber floor, and subsequent investigation showed them to be dead. In a profession well-known for its high risk of hypothermia, they had died of *hyper*thermia (heat-stroke).

This accident has been described at some length because, more than any other single incident, it has been picked up by the media and by others as epitomizing the way in which diving takes its practitioners close to, and sometimes across, the threshold of the known and the familiar. Indeed, one legal text on offshore law cites it as 'a good example' of the fact that some fatalities 'occur in situations where very little can be done in any event, or in a totally unforeseen manner' and as an illustration of how 'divers are operating at the limits of modern underwater and medical technology.'[40] Such a conclusion might indeed seem to be justified by the bare facts recounted above, as by a point made by the Sherriff at the Fatal Accident Inquiry which ensued:

> Although all the components in this tragedy were known to science – particularly to heat physiologists – their possible conjunction in a compression chamber had never previously been recorded or, so far as the evidence showed, foreseen by those engaged in practice and research throughout the long history of deep diving.

The affair assumes a rather different complexion, however, when some of the other findings of the Inquiry are taken into account. The pressure gauge in question was fitted to the main deck decompression control panel but could be made to monitor pressure in any other part of the complex by the opening and closing of valves. Its use was preferred to that of the available fixed gauges for each individual part of the complex because of its finer calibration. Thus it was possible for the supervisor to think he was monitoring one unit while actually monitoring another, an error which the Fatal Accident Inquiry found to be substantially responsible for the accident. Moreover, and quite apart from the inherent deficiencies of a system that has a gauge 'capable of being used to indicate pressure in a compartment other than that for which it was clearly labelled', it was held that this particular layout was especially dangerous since there was a prominent notice saying 'MAIN LOCK' (i.e., main compression chamber) prominently displayed over the gauge in question (Case 22). High technology, perhaps, but this was a comparatively straightforward design issue coupled with operational error, which, it should be remembered is amenable to proper operational rules. In the circumstances, it is not perhaps surprising that this incident should have provoked one experienced diver to write to the authorities complaining about what he called 'the scandal of the legalized manslaughter which has been taking place in the North Sea since the search for oil began'.

Most of the other diving fatalities in the series can be described more briefly, if in terms which must necessarily appear somewhat sanguine. Four divers were killed during operations, not at some incredible depth, but on or very close to the surface. For reasons which were never established, one man attempted to cut himself free from his lifeline and to slip his weight-belt at a depth of 15 feet. The fact that he was wearing the latter underneath, rather than on top of, the jock-strap of his breathing apparatus probably meant that, having success-fully freed himself from the line, the weights slipped down to hang beneath his crotch and pull him farther down into the water (Case 5). Another diver was thought to have broken his ribs against some surface obstruction and drowned in the course of attempts to pull him from the water, although he had previously been observed to be in some difficulty with his breathing equipment, and some faults were indeed subsequently found in it (Case 11). A company and its diving supervisor were prosecuted in 1977 following the death of two divers who had been allowed to enter the turbulent seas around anchor bolsters at night with an inadequately secured 'buddy-line', no stand-by diver and no effective arrangements to secure compliance with regulations (Case 34).

The last 'surface' death inolved a diver who, having experienced problems in getting back into the diving basket suspended on the rough surface, allowed himself to drift away, apparently unperturbed, in order to be picked up by the stand-by boat. The latter had some difficulty in manoeuvring close enough to effect the rescue and was thought to have either run him down or dragged him into the propellor screws by his trailing 'buddy-line'. Had the slings holding the diving basket to the main crane wire been of sufficient length, the basket could have been lowered through the turbulent surface interface to a depth where re-entry difficulties would not have been experienced in the first place.[41] The Sherriff's conclusion at the subsequent inquiry was pointed:

> I also determine that the death was contributed to by the inefficiency of the method used by the rig to recover divers from the water, and by the fact that the stand-by vessel was unsuitable for rescue purposes, and her crew were inadequately trained in life saving. (Case 37)

Apart from one extraordinary incident in which a medical practitioner misdiagnosed pneumothorax as pneumonia and persisted in this opinion despite other well-informed advice to the contrary

(Case 9), the remaining three fatalities came closer to diving's traditional image. Again, however, closer inspection revealed that although advanced technology and the frontiers of human endurance were certainly involved, there were also, and crucially, much more familiar factors at work. Such, for example, would seem to have been the case in the instance of a fatal accident which took place at a depth of 480 feet, when a diver exited from the submersible compression chamber having probably knocked the main gas supply valve closed with his 'bale-out bottle'. The supply in his 75-foot umbilical was sufficient to allow him to leave but not to re-enter the bell (Case 25). Similarly, a saturation dive to 335 feet, using a breathing system so advanced that there was no establishment in Britain competent to undertake its testing, ended in tragedy when the diver working outside the bell closed the gas supply valve on his helmet. Whether this was because he went too far in trying to reduce discomfort arising from water that had previously been noticed in the gas supply system or because he acted irrationally following a build-up of carbon dioxide in his helmet was never established. As a factory inspector noted at the time, however, while there was no doubt that the company was 'working at the frontiers of knowledge in the mechanics of this breathing system', it was a system which 'could not be used under the Factories Act, without special exemption from the Chief Inspector of Factories' (Case 3).

The final case demonstrates very clearly that we should be careful not to be blinded by the high drama of North Sea diving accidents. Early in 1976 two divers entered a diving bell in order to carry out the inspection and subsequent replacement of the flowlines from a well head. The vessel from which this operation was to be conducted had originally been designed as an oil-rig supply boat, but some six months earlier a diving unit had been fitted to its stern. The diving bell in question had also been modified because preliminary trials had shown that surface conditions and associated ship movement caused a 'bouncing' effect when it was lowered to the sea bed. In an attempt to eliminate this uncomfortable and potentially dangerous effect, the dropweight which had originally been attached to the bell itself, thereby giving it straightforward negative buoyancy, had been removed. In its place, a basket-type dropweight system had been introduced, whereby the necessary weight could be suspended some distance *below* the bell, the latter now being held in its position above the basket by its own positive buoyancy and by two chains anchoring it to the basket beneath (see figure 2). These chains were attached to release mechanisms on the bell in order to allow the basket to be jettisoned in an emergency. Once the bell weight had been placed on the sea bed and the bell's positive buoyancy had come into action, the bouncing effect could be eliminated by paying out some slack on the main hoist wire from the vessel on the surface and by engaging a heave compensator which had been introduced at the same time. Theoretically, the bell would then remain buoyant, suspended above the weights and unaffected by the ship's movement.

— umbilical

— access hatch

— cable

— ballast

Figure 2
Diving bell with basket dropweight
(artist's impression)

On the afternoon in question, the two divers were lowered in this bell to a depth of 256 feet, where, with the basket weight resting on the mud-covered bottom, the bell was pressurized to an appropriate level and the door opened. One of the two men then proceeded to examine the relevant pipelines, but found that the length of his umbilical was not sufficient to permit the necessary work to be carried out. Accordingly, it was decided to raise the bell and its basket some 20 feet, and, by warping the ship some distance to starboard, relocate the underwater system in closer proximity to the task. The diver who had been working outside returned to the bell and stood on the outer bottom door with his head in the bell's atmosphere while preparations for this manoeuvre were made. When the lift began, the basket dropweight was suddenly released and the bell, with the bottom door still open, shot to the surface in the space of less than a minute. During the ascent, the second diver successfully pulled his colleague inside, but was unable to secure a seal on the open door. The effects of such rapid decompression killed his companion and, some ten weeks later, he himself was still paralysed from the chest downwards (Case 26).

This episode contains several of the ingredients necessary for a good newspaper story about the dangers of the North Sea – diving, a dramatic mode of death and the image of an industry using its ingenuity to overcome difficult operating conditions. Moreover, the fact that a system operating on the same principle had already been in use for some years in the Mediterranean and elsewhere lends credence to the view that North Sea accidents somehow reflect unique operational problems. In the same context, it is significant that in the aftermath of this accident, diving companies were warned that while a supply boat of the size in question might be adequate for calmer water conditions, it could be too lively a vessel to support a saturation diving system in comfort and safety in the North Sea.

None of this can be gainsaid, but one or two facts can be added. Thus, for example, expert investigation of the reasons for the basket weight's becoming detached revealed that one of the release mechanisms to which the anchoring chains were attached had broken under the strain of the lift. Capable of withstanding a force of about 2 tons, each of the attachments had been subjected to a force possibly as high as 45 tons, since the basket had to overcome the effects of suction before lifting clear of the muddy sea bed. The failure of the release mechanism, the report laconically concluded, 'was due to the design being inadequate to withstand the forces involved'. Indeed, while there was some doubt as to whether the crane in use at the time would even be capable of producing the total lift needed to raise the system in the prevailing circumstances, 'the motion of the ship stern in a 6-foot swell would produce forces in the lifting cable large enough to break either the suction force, the main lifting cable, the crane, the lifting lug, the cables attached to the ballast tray or the release mechanism, whichever is the weaker.' If it was only with the benefit of technically informed hindsight that the failure which occurred in these circumstances could be described as 'not unexpected', the disaster which struck had not been totally unheralded. During operational trials some three months earlier, the chain slings holding the basket on the new system had parted, 'due to some unknown force'. New slings had subsequently been designed and produced using specially tested wire strops, but we do not know whether the vulnerability of the system's other parts to such unknown forces was taken into account. Common sense, not to mention good engineering practice, would suggest that they should have been.

In diving, then, as in other dangerous offshore occupations, mundane design faults, human error and unsafe working procedures seem to play as significant a role in accident causation as do the exigencies of high technology, adverse operating conditions and limited human physiology.[42] These latter images, with their possible implications of only a very limited role for law and its enforcement in the prevention of accidents, have to be rejected. On the evidence adduced by this research, the real problems of safety in the North Sea are often the same as those encountered onshore and should be treated accordingly. It is all too easy for those caught up at any level in a particular industry (or for those charged with its regulation)[43] to become mesmerized by its unique features and reconciled to its reputedly special risks.[44] The oil industry would not be the first to stand in need of a gentle reminder in this respect. The conclusions of a survey of fatal accidents in another 'especially' dangerous industry could, *mutatis mutandis*, be applied to the offshore industry without too much difficulty:

> examination of these one hundred accidents and of the national pattern of fatal accidents suggests that we must isolate and give publicity to the important and identifiable groups of more mundane accidents which cause the greater number of preventable deaths in the construction industry. . . .
> None of the accidents described in this group caused more than a brief flurry of interest in the local papers. The men involved were construction workers going about their normal work; they were not

working at the frontiers of technology; they were simply picked off one at a time. . . . There were very few freakish accidents, or accidents which would have surprised someone who had worked or inspected in the industry for several years. The vast majority of these accidents can be prevented by the competent exercise of normal professional skills, by adequate training and supervision, and by the establishment of safe systems of work.[45]

Oil Pressure

If the very real dangers associated with offshore employment cannot be traced primarily to the unique exigencies of advanced technology, harsh climatic conditions or operations which entail the constant testing of the physiological limits of mankind, to what can they be attributed? One possible culprit, a factor that has reared its head on more than one occasion in the course of preceding pages, is, of course, human error. Some support for the indictment of this all too persistent offender can, moreover, be marshalled from the one series of accident summaries supplied replete with departmental comments on the relevant incidents.[46] Out of sixty-six fatal, serious or dangerous accidents which fell inside the Department's purview and upon which some comment was made in these summaries, forty-two, or 64 per cent, were laid wholly or in part at the door of human error. How this figure would be affected if comments on the remaining accidents were available or if access had been obtained to a more extensive sequence of complete summaries cannot be determined, however.[47]

But what does 'human error' mean? Does it carry the inescapable implication that we must simply be reconciled to the consequences, unsavoury as they may be, of human fallibility? To an extent, such is obviously the case. But the comments made by the Norwegian Royal Commission of Inquiry into the potentially catastrophic blow-out on the Ekofisk Bravo platform in July 1977, an investigation which ascribed the accident largely to human error, enter some interesting qualifications on this score:

> Human beings err from time to time, and it must be anticipated that technological components may fail too. Even if all technological components function perfectly, that does not automatically make it correct to name human error as the cause of an accident. Each and every technological system must be composed to allow for the fact that it will be operated by humans, and humans may err. The term 'human error' should be reserved for those cases where the sum of

human actions exceeds reasonably acceptable limits. The administrative system must be attuned to the human and technological
material at disposal, and must be designed so that it detects indica-
tions of human error or technological failure and effects the counter-
measures necessary.[48]

Whether the British Department of Energy is more (or less) promis-
cuous than the above report in its use of the term 'human error' is not
something that can be established, though I was quite rightly
cautioned by one senior official to refrain from jumping to the conclusion that use of the term, whether in interviews or elsewhere, implied
violation of any legal code. Having acknowledged this, however, it
must also be said that the implementation of the safe working practices
and 'administrative systems' that can minimize the risks of human
error, as envisaged in the Norwegian report already cited, is precisely
one of the points at which law can intervene most beneficially with
regard to general standards of safety at work.[49] Moreover, and quite
regardless of law, it is obviously true that some situations are more
conducive than others to errors of judgement, and we can therefore
ask to what extent the North Sea oil industry is prone to such circumstances. In attempting to answer this question, the other image associated with offshore risks, that of an industry which is relentless in
the demands that it makes of its workers and ruthless in its pursuit of
economic objectives even at the cost of safety, will come under rather
closer scrutiny.
     According to one of the industry's most frequent commentators,
'fatigue, cold, hunger and, not surprisingly, boredom are major factors
contributing to the accident rate.' While the first three of these are dealt
with adequately by 'working conditions, working clothes with a
degree of insulation, and excellent food', he insists, boredom is much
harder to combat as 'the crews spend a great deal of time just standing
around.'[50] This rather leisurely image of the bored worker whose
attention momentarily wanders, to his own or others' subsequent
detriment, is not quite in line with the impression of the offshore work
situation gained in the course of this research. While a substantial
number of the workers spoken to did indeed concede, and readily, that
carelessness or lapses in concentration are major contributors to
offshore risks, boredom through inactivity was not cited as their
underlying cause. On the contrary, the picture painted was one of
hard, long and continuous labour – as one roustabout put it, 'It's hard
work and intense, it's kind of, like, pressure; you just eat, sleep and

work.' Indeed, far from 'just standing around', it was suggested, if you have nothing to do at a particular time, you must look busy.

With a working day of twelve hours for every day of a tour of duty which, in the case of most non-management personnel in the North Sea, lasts for two weeks, fatigue and other problems associated with the duration of labour should not perhaps be so readily discounted, even if we are not here talking of the 'physiological limits' as that term is applied, say, in the context of diving. As a Health and Safety Executive official commented, there is certainly a question to be asked about 'how far this kind of intensive working, day after day, for a day which is 50 per cent longer than the norm, is compatible with safety'. Similar queries have been raised by the Norwegian risk-assessment study referred to above and by the International Labour Office's review of safety problems in the offshore petroleum industry.[51] According to the latter, for example, 'such a working schedule combined with extreme environmental conditions and extended isolation from families and friends, can have serious effects on offshore personnel morale, fatigue and, in the end, safety.'[52]

Such research as has been carried out on this subject apparently suggests that, at most, there is a slight tendency for accidents to increase both at the beginning and at the end of a tour, with a similar trend occurring just before and just after shift changes.[53] As far as the present project is concerned, however, it was not possible to obtain sufficient or systematic data which would confirm or refute these reported findings, though one of the largest operating companies did reveal that an increase in drilling accidents around the time of shift changes was the *only* pattern to be thrown up by its accident statistics. Worth adding, perhaps, is the fact that in 1978, this same company was still only starting to computerize its accident reports, so that 'in the fullness of time we can expect to extract . . . all sorts of odd things we couldn't at the moment do, unless we went through a laborious manual process.' Such a gap between the level of technology employed in the processing of accident data and, for example, the sophisticated banks of computers used in controlling the operations of a production platform possibly speaks for itself.

Although the problem of fatigue was not something that this project could investigate in any really systematic fashion, some of those interviewed did have interesting comments and opinions to offer on the connection between safety and work schedules. Several workers, for example, did indeed claim that tiredness plays a substantial part in causing accidents, and that the risk is therefore greatest towards the end of a twelve-hour shift. Interestingly, however, they added that the problem of fatigue is at its height during the early stages of a tour of duty, before people have become accustomed (or reaccustomed) to the routine of working offshore. This possibility would be in keeping with the almost universally agreed view that lack of experience, coupled with a high rate of staff turnover (as high as 40 per cent per two-week shift, according to the reported estimate of one drilling company), is a major cause of offshore accidents.[54] To this it may be added that the problem of acclimatization is certainly exacerbated on some installations by a level of noise which necessitates the dispensing of drugs such as Mogadon to enable workers to sleep when off-duty, again particularly during their first few days on board. Some personal experience confirmed the problem, though not the remedy. It was also suggested by one rig medic that the question of fatigue in the early stages of a tour is sometimes not helped by inadequate onshore accommodation arrangements prior to the transfer of personnel offshore.

The inexperienced worker, unaccustomed to long shifts, may well be at special risk offshore. But everyone's vulnerability is increased if, on top of protracted hours of work and long tours of duty, he has to work under intense pressure. Here we move closer to the traditional image of the oil industry as one which imposes just such pressure. Drillers, it was said, frequently cause accidents by their 'impatience in trying to better previous shift times' and will 'put safety by the board, if it means getting the job done quicker'. Toolpushers were similarly described by one safety officer as having a kind of inbuilt desire to maintain drilling progress at all costs, and, as a roughneck explained, are not themselves immune from pressure:

> You get the odd one who is genuinely concerned about the men's welfare, but they are few and far betwen. They are obviously under a lot of pressure from the company they are employed by, looking for figures. I mean it costs a lot of money to operate an oil or production platform, so there is pressure all the way down the line.

At various points in this enquiry, other ways in which such 'pressure all the way down the line' might adversely affect safety standards became apparent. Supply boats, for example, were said to come alongside even in adverse weather conditions, not least because of a very tight competitive market in the supply business. As one union official put it, 'It's the old commercial pressures put on skippers and on rig masters – you know, "We need that steel. We've got to keep drilling no matter what the weather conditions are." ' The head of one well-known company's safety department confirmed that there is, indeed, considerable pressure on supply contractors in this respect, while a senior official at the Department of Energy specifically chose supply boats as his example of how such pressure could sometimes militate against safety. When asked whether he thought ordinary commercial pressure, combined with the scramble to get oil ashore as quickly as possible, might have anything to do with accidents, he replied:

> Yes, I think, being quite honest, I think it might do. It becomes a question of someone's judgement when it is safe to do a thing or is it not safe to do a thing, which is very hard to pin down. . . . It has to be the man on the spot that's doing it, and it maybe [happens] that people will sometimes try and offload a supply boat in weather which is perhaps a little bit worse than they should do. And then you get someone being caught by a load swinging or a mooring rope going. One case was of a mooring wire breaking and hitting somebody. If the weather had not been quite so rough, this might not have happened.

Such a comment provides a sombre gloss on the catalogue of crane accidents discussed earlier, as on the general problem of the pressure under which responsible people have to take decisions which may have implications for safety. Thus, for example, an instrument engineer claimed in interview that production management had ordered him to close down gas sensors because they were 'too sensitive', and that, as a consequence, these vital safety mechanisms had been inoperative for a period of twelve hours. A safety officer conceded that although he would probably get his own way at the time if he really objected to some major aspect of an operation, he would almost certainly encounter subsequent problems 'back on the beach'. His point was put more forcibly by a union official, who expressed extreme scepticism as to whether safety officers, despite their protestations, would ever in fact 'buck production' to the point of insisting on a shut-down which might cost the company thousands of dollars a day. Similarly, from direct experience of several confrontations and other, less fraught discussions between safety officers and operational management on board installations, it became fairly clear to the writer that, despite the safety propaganda so prominently displayed in mess-rooms and elsewhere, 'Safety first' was not always an immediately acceptable operational priority. Nor, according to some well-informed scources, are safety officers the only ones to face such dilemmas:

Where a diving superintendent may say, 'Look, conditions are just not right for my lads to go down today' or 'They have been down too long over the last two or three days and they need to be rested up', the rig master can say to him, 'Well, I've got pressures; I've got to get these jobs done. If your company won't do it, we'll soon get another company to do it'. . . . And you have got fairly experienced men in these positions, but their professional judgement is far too often, I think, clouded by the commercial pressures.

If those who have to take decisions are under these kinds of pressure, what of those who have to execute their orders? Although I was assured that the days of the 'real cowboy companies' had now passed, the experiences and opinions related in interview often created a strong impression that the general offshore atmosphere is still not one in which objections are readily countenanced. One diver put it at its enigmatically mildest when he explained, 'No one is going to make you do an extra dive when you've done your number of hours already; but you know you've got to go.' More starkly, a roughneck asserted that there is, even now, fear among many offshore workers that any objection will result in their employment being terminated. According to another rig worker, you might get away with it by sitting down beside the rig superintendent during a meal and saying, 'Isn't this a breach?' but confrontation, on any score, has only one result:

It doesn't even [have to] concern the safety of the rig. The rig I was last working on was moving to Canada, and there was trouble about the bonus. They needed the crew because they were only going to drill for three months in Canada, and so they needed trained people to get the job done in time. And the trouble was . . . they gave the management an ultimatum, so they said, 'OK, you are fired.' And they had a helicopter out the next morning for them.

While an incident such as this may be a rare relic of the 'bad old days' of ten or fifteen years ago, and may be even rarer when safety is the issue at stake, there is no doubt that some middle-range and senior personnel in the oil industry do still exude the sense of toughness that has won that industry its ruthless image. As we shall see at a later stage, for example, intolerance of legal contraints and impatience with regulatory codes seen as pampering the workforce are by no means entirely unknown.[55] Certainly, little by way of the latter might be expected from a Texan rig superintendent who told me that, as far as he was concerned, the offshore worker 'is paid his money and can take his chance'. Even when the pressure is not couched in quite such stark terms, it can still be there and can still play its part in creating unsafe situations:

> The Commission has not found any attitude or circumstances on the part of the licensee to indicate that for economic reasons the work-over operations were speeded up or otherwise planned in such a manner as might be detrimental to safety. Nevertheless, the Commission can naturally not exclude completely the possibility that individual persons or the crew as a whole may have felt they were under pressure to complete the work quickly, since it was so greatly delayed. This probably explains the long hours worked by the Otis operator. That he worked for a 30-hour period without sleep was not justifiable.[56]

If 'pressure all the way down the line', whether in terms of specific decisions or in terms of the more general ethos alluded to above, cannot be discounted as a factor in offshore risks, it becomes germane to ask where this line stops if it is followed not downwards but upwards. Clearly, the answer cannot simply be 'at the level of on-board management', since, as we have already seen, such personnel may themselves be under pressure from their shore-based companies. Moreover, as often as not, the latter will merely be contractors who are therefore vulnerable to further pressure from operators or licensees. As one respondent explained, the days when excess demand for their services could conceivably allow a contractor to 'buck the operator' have gone, the latter now being in a position to require whatever is needed, under virtual threat of hiring another contractor. This distribution of market power could, of course, mean that an operator has useful sanctions to deploy against contractors who do not adhere to prescribed standards of safety, but it also means that pressures of other kinds can be applied. While variable accounts were offered of how effectively such power is exercised in the first of these respects, with regard to the second it should be remembered that speed and the control of costs are of the essence, at least from the companies' point of view. Guy Arnold has put the general point very cogently:

> The answer, as always, comes back to the economics of oil – and these should never be overlooked. More controls and supervision, such as further imposition of government safety standards, can slow down the exploitation effort. In an industry where huge worldwide shortages may develop in the next few years thus causing the price to escalate correspondingly, no one wants any slow-down to take place – no matter what extra risks this attitude entails. That is the nub of

the oil company approach: profit. That, at any rate, is one side of the picture.[57]

Just how far such considerations may impinge upon decisions which concern safety is not easily determined. Not suprisingly, most of the shore-based safety managers interviewed were somewhat reticent about the extent to which any of the plans and policies emanating from their departments would be subjected to rigorous scrutiny in terms of cost analysis. Thus while several reports of 'in-house' surveys purportedly showing that the two weeks on/two weeks off system involves a higher accident rate than does the one week on/one week off rotation were encountered, it was not possible to confirm the existence of these studies, their alleged findings or the veracity of additional claims that the second alternative had subsequently been rejected on grounds of cost. Similarly, any suspicion that the replacement of two twelve-hour shifts by three eight-hour ones might have foundered on economic grounds was not substantiated, although it might seem slightly disingenuous for an expert in the industry (which has hitherto underestimated accommodation requirements)[58] to claim that the significant advantage of the twelve-hour system is its reduction in the number of people at risk.[59] Equally, I was assured by one official that it would not be economically worthwhile for the industry to neglect safety, particularly since the costs, as he put it, 'are insignificant when compared with profits'.

Against this comforting view, however, must be set other evidence and opinions suggesting that costs and similar considerations are not quite as irrelevant to the safety calculus. One of the Norwegian risk-assessment study's conclusions, for example, was that while some remedial measures can be introduced without any cost increases, such was not the case for all the steps recommended. 'As there is a limited amount of money available for risk-reducing measures', it went on, 'the aim should be to find the areas in which the available money has the largest risk-reducing effect.'[60] Such a proposal would scarcely seem to reflect a situation in which cost is an insignificant factor in the allocation of priorities with regard to safety – nor, to be more specific, would the continued use of dynamically positioned vessels for diving operations, a system described as 'cheaper and more convenient than traditional multi-point mooring systems', despite a safety record which could provoke an uncharacteristic threat of a complete ban from the British Department of Energy.[61] The more general point has again been cogently expressed by one Certifying Authority:

it would be imprudent to ignore the economic and political pressures upon the operator for continuity of oil production. It is not clear whether there are adequate safeguards built into the management system which would enable safety to be given equal prominence at senior-management level in cases of a conflict of interest.[62]

With such doubts I can only agree. What seems to be being hinted at, however, is that the line down which pressure may be exerted does not simply run from a single point of termination somewhere in the economics of the industry. Political pressures are also called into question. Thus the image of an industry within which safety may fall prey to a particularly harsh set of internal economic 'realities' begins to merge with that of an industry whose safety record, at the end of the day, cannot be discussed in isolation from the position held by oil in a broader political nexus. Something of this kind seems to have been in the minds of the directors of the Norwegian Petroleum Directorate when they observed, in their report for 1978, that although safety measures involve considerable costs and must therefore be weighed against the benefits which may accrue, 'in the last instance, such matters will often be a question of opinions based on the dominant values of the society.'[63]

Whether the 'dominant values' expressed in the application of British law to the safety of offshore operations are superordinate to, subservient or even divergent from the values, interests and policies of the oil industry is an issue to which we shall return at many points in subsequent chapters. More specifically, while it may not be possible to lodge a categorical indictment against pressure for speed as the main proximate cause of accidents on installations, there is no doubt about its responsibility in another crucial respect – the failure of British safety legislation and administration to keep pace with offshore developments. To understand how and why the latter should have come to outstrip the regulatory response, however, it is necessary to depart temporarily from the question of safety and to examine how speedy development became the dominant concern, not just for the industry but also for successive British Governments. Against this background, we will then subsequently return to the laggardly nature of offshore safety legislation and to the growth of a regulatory regime that, preoccupied with the major risks attendant upon rapid technological and operational progress, paid relatively little attention to the more mundane risks which this chapter has shown to underlie the British North Sea's poor safety record to date.

80 Images of Danger

Notes and References

1. In the Fastnet yacht race fifteen people died, as compared with four deaths reported on or around oil installations in the British sector during 1978.
2. All headlines quoted are from the relevant newspapers published on 9 August 1979.
3. *Observer*, 12 August 1979.
4. Adrian Hamilton, *North Sea Impact: Offshore Oil and the British Economy*, London, International Institute for Economic Research, 1978, p. 69.
5. Guy Arnold, *Britain's Oil*, London, Hamish Hamilton, 1978, p. 9.
6. 'Offshore Europe '79', special supplement in the *Scotsman*, 3 September 1979, p. xxi.
7. *Aberdeen Press and Journal*, 17 July 1979.
8. This story is recounted by Dr J. Kitchen, who collaborated with the author during a preliminary investigation to assess the feasibility of conducting research in this area.
9. *Guardian*, 21 June 1977; see above, p. 15.
10. *Guardian*, 10 August 1979.
11. *Aberdeen Press and Journal*, 10 August 1979.
12. *Guardian*, 19 September 1979.
13. This play, by John McGrath, was performed throughout the crofting counties and the south of Scotland from 1973 onwards by the 7:84 Theatre Company.
14. *Guardian*, 20 September 1979; the remission of large profits to parent companies abroad and the payment of foreign loans, another factor behind the deficit, both reflect the high preponderance of foreign interests in the North Sea. For some details and an explanation, see chapter 4 below.
15. J. K. Galbraith, 'Oil: a Solution', *New York Review of Books*, 27 September 1979.
16. Charles F. Doran, *Myth, Oil and Politics*, New York, Free Press, 1977, p. 48.
17. British Medical Association: Scottish Council Report, *The Medical Implications of Oil Related Industry*, Edinburgh, 1975, p. 4.
18. See, for example, Mervyn Jones and Fay Godwin, *The Oil Rush*, London, Quartet Books, 1975, p. 59; Jonathan Kitchen, *Labour Law and Offshore Oil*, London, Croom Helm, 1977, p. 129; International Labour Office, *Safety Problems in the Offshore Petroleum Industry*, Geneva, ILO, 1978, p. 48.
19. Kitchen, *Labour Law and Offshore Oil*, p. 172.
20. Jones and Godwin, *The Oil Rush*, pp. 37ff.
21. *Scotsman*, 28 July 1979.
22. Calculated from unpublished quarterly accident summaries supplied by the Department of Energy; see pp. 50 ff. below.

23. According to one Department of Energy inspector, this is more or less the view adhered to by several of his colleagues.

24. According to this report, 25 per cent of accidents result from falls; over-exertion, being struck by objects and being caught between objects each account for a further 17 per cent. Royal Norwegian Council for Scientific and Industrial Research, *Risk Assessment: A Study of Risk Levels Within Norwegian Offshore Petroleum Activities* (hereafter *Risk Assessment*), Report No. 26–27/2, Trondheim, 1979, p. 91.

25. Occupational Safety and Health Administration, *Selected Occupational Fatalities Related to Oil/Gas Well Drilling Rigs*, Washington, US Department of Labour, 1980, p. 5.

26. An attempt at such quantification, based on the full Department of Energy files, was made in the course of the Burgoyne Committee's inquiries into offshore safety. Covering accidents and dangerous occurrences which took place between the beginning of 1974 and September 1979, the results showed structural and mechanical failure to have been responsible in approximately 33 per cent of the incidents, while impacts on persons or structures by moving objects accounted for a further 22 per cent. Falls only came to around 14 per cent. *Offshore Safety*, Cmmd. 7841 (hereafter the Burgoyne Report), London, HMSO, 1980, pp. 116–7. The discrepancy may be accounted for either by the inclusion of dangerous occurrences or by the fact that the survey was able to reclassify accidents. However, the contention in this chapter is that there is more to be gained from detailed descriptive material pertaining to a relatively small number of accidents than from classification and quantification on a wider scale, an exercise which must necessarily be somewhat arbitrary.

27. Occupational Safety and Health Administration, *Selected Occupational Fatalities Related to Oil/Gas Well Drilling Rigs*, pp. 1, 2, 49.

28. For another use of this method, see Health and Safety Executive, *One Hundred Fatal Accidents in Construction*, London, HMSO, 1978.

29. These files are held in the offices of the Procurators Fiscal who, apart from duties relating to prosecutions in Scotland, are also responsible for investigating accidental or sudden deaths. This they do with the assistance of the police and the relevant enforcement agency, where appropriate. Hence the files contain reports from the Department of Energy, as well as other expert reports, witness statements, etc. For a further discussion of the Procurator Fiscal's role (and difficulties) with regard to the North Sea, see chapter 7 below.

30. Some cases were still under investigation and hence were unavailable, while others which had taken place in the English sector would obviously not be held in Scotland.

31. See pp. 310 ff., below. Apart from the standardization of distance measurements, accident details are as stated in the relevant reports.

32. See, for example, Burgoyne Report, p. 10.

33. See, for example, *Development of the Oil and Gas Resources of the United Kingdom*, London, HMSO, 1977, p. 15.
34. The Norwegian Petroleum Directorate, for example, attributes part of the steep rise in accidents on the Norwegian Continental Shelf during that year to 'the many simultaneous working operations, often narrow working places versus a large number of personnel and often changing weather conditions – together with extensive construction work'. English extract from *Oljedirektoratet: Arsberetning*, Stavanger, 1978. See also Burgoyne Report, p. 11.
35. See chapter 6 below.
36. *Oilman*, 2 April 1977.
37. *Risk Assessment*, p. 149.
38. *Oilman*, 2 April 1977.
39. *Edinburgh Evening News*, 8 August 1979.
40. Kitchen, *Labour Law and Offshore Oil*, p. 129.
41. Because of the effects of immersion on lifting gear (see above, p. 56), operators are normally forbidden to allow crane blocks to become submerged. Hence, adherence to this rule, coupled with the shortness of the slings, meant that the basket could not be lowered through the surface interface.
42. Six further accidents in the series did not lend themselves readily to classification under the headings used here. Apart from one fatality involving a helicopter, however, they seem to fit the pattern which has been described. Two men working on barges were killed in the course of anchor operations, and two others fell to their deaths when they were not wearing life-lines. The sixth case involved a pipe fitter who fell into the sea when the grating upon which he was standing gave way. The grating had not been securely clipped in place to prevent lateral movement (Cases 10, 13, 24, 28, 33, 38).
43. For a discussion of how the Department of Energy became preoccupied with the 'unique' aspects of the offshore industry, see chapter 5 below.
44. At a meeting of experts convened by the International Labour Office, for example, 'the employer experts stated that they were not convinced that many offshore safety and health hazards were similar to those on land.' International Labour Office, *Safety Problems in the Offshore Petroleum Industry*, p. 98. In fairness, it should be added that the differences were accepted as being such as to necessitate more stringent safeguards in the offshore context.
45. Health and Safety Executive, *One Hundred Fatal Accidents in Construction*, p. 12. The Norwegian Petroleum Directorate has also listed basic training, more frequent control, better 'housekeeping' and better organization, planning and execution of the job as the best means of reducing the number of injuries occurring in the Norwegian sector. *Oljedirektoratet: Arsberetning*, p. 3.
46. This series covers the first nine months of 1976.

47. Although all summaries from 1974 to 1978 were made available, apart from the series mentioned above, all had the column headed 'Comments' cut off, and all names, dates, locations, etc. obliterated.

48. *Report from the Commission of Inquiry into the Uncontrolled Blowout on Ekofisk Bravo, 22nd April, 1977* (preliminary edition), Oslo, 1977, p. 109.

49. Such, for example, is the general ethos of Secion 2 of the Health and Safety at Work, etc. Act 1974 (c. 37).

50. *Oilman*, 11 August 1979.

51. *Risk Assessment*, p. 138; International Labour Office, *Safety Problems in the Offshore Petroleum Industry*, p. 91.

52. ibid., p. 19.

53. ibid., pp. 19 and 91.

54. One safety manager, for example, pointed out that the main problem lay in these two factors, which affect the contracting companies in particular.

55. See chapter 7 below.

56. *Report from Commission of Inquiry, etc.*, p. 107.

57. Arnold, *Britain's Oil*, p. 27.

58. *Development of the Oil and Gas Resources of the United Kingdom*, London, HMSO, 1977, p. 15; 1978, p. 15.

59. International Labour Office, *Safety Problems in the Offshore Petroleum Industry*, p. 91.

60. Royal Norwegian Council for Scientific and Industrial Research, *Overall Risk Assessment*, Report 26–7/1, Draft 2, 1978, p. 128.

61. *Oilman*, 16 June 1979.

62. Burgoyne Report, p. 151.

63. *Oljedirektoratet: Arsberetning*, p. 80.

4

The Political Economy of Speed

My people think money
And talk weather. Oil-rigs lull their future
On single acquisitive stems. Silence
Has shoaled into the trawlers' echo-sounders.

The ground we kept our ear to for so long
Is flayed or calloused, and its entrails
Tented by an impious augury.
Our island is full of comfortless noises.
Seamus Heaney

To many students of the sociological processes involved in legislation, law enforcement and the like, the title of this chapter may seem redolent more of a work on drug abuse than on North Sea oil. While the analogy is tempting – oil certainly did 'lull features' even if it has not quite turned out to be the economic stimulant some had hoped – it is not, however, my intention to pursue such a colourful theme. Rather, my purpose in this chapter is the comparatively straightforward one of exploring the concatenation of forces which gave rise to the dramatic speed with which the mineral resources of the British Continental Shelf were developed. In so doing, I shall map out in a little more detail than heretofore the policies pursued by successive Governments, and will suggest that their commitment to speed in this context entailed some important consequences for the nature of the relationship between the British state and the offshore oil industry. Against this background, chapter 5 will examine how this same commitment to speed and its relational implications permeated the growth of a special regulatory regime for controlling the safety of operations in the North Sea.

In order to undertake this task, it is necessary to recall that, as

The extract from Seamus Heaney's 'Sybil' is reproduced by kind permission of Faber and Faber, and in the USA by permission of Farrar, Straus and Giroux.

suggested at the beginning of this book, one of the issues thrown into particularly sharp relief by the history of offshore developments is the need for analysis to be couched in terms which take cognizance of a totality that is wider than the nation state. Vital to the process of accumulation in the West, central to its relationship with that part of the world upon which it remains substantially dependent for supplies, of paramount strategic significance in East–West relations and of crucial relevance to the predicament of underdeveloped countries, oil has perhaps become the international commodity *par excellence*. As a consequence, the opening up of the British Continental Shelf has to be set against the backdrop of developments within the world economic system and, no less important, against the background of the political and economic events which have shaped the pattern of oil supply, consumption and pricing since the Second World War. Together, these factors may be described as the external dynamic in a process sometimes referred to as 'combined and uneven development'.

On the other hand, and as this term itself implies, it is also important to realize that development, even within the world system of capitalism, is uneven and that movements within the international context cannot therefore tell the whole story with regard to Britain's oil. The nation state, as David Purdy observes, should not be seen as 'merely registering in its internal motions forces at work on a global scale'.[1] In consequence, it also becomes imperative to locate analysis of the burgeoning offshore industry within some discussion of the historically specific and even unique features of Britain during the operative period, features which provide one vital part of the key to the oil policies adopted by successive Governments over the past twenty years. It is only when the interplay between external forces and internal exigencies is grasped that we can begin to understand the political economy of speed.

The Sixties and Gas

The complex combination of external and indigenous forces which conjoined to perpetuate the decline of the British economy after 1945 and to delay the institution of 'necessary' remedial measures by the British economic and political establishments has been discussed by a number of recent writers and need not be rehearsed in detail here. In broad outline, however, there is general agreement that although the malaise had much deeper historical roots, the purely temporary appa-

rent reprieve of the post-war era served only to accentuate the underlying downwards trajectory. Some of the major factors involved have
been admirably summarized by Bob Jessop in a recent paper on the
state in post-war Britain:

> First, Britain was characterized by the separation of financial and
> industrial capital and the economic and political hegemony of the
> former to the long-term detriment of the latter. Secondly, owing to
> the possibilities of a retreat into imperial trade and investment and to
> the defensive strength of key sections of the labour force, industrial
> capital in Britain became increasingly less competitive in international
> markets. Thirdly, despite the major decline in Britain's economic and
> political position in the international order by 1945, a concerted
> attempt was made to restore sterling and empire to a central position
> in the new world system. And, finally, within the framework of the
> post-war settlement between capital and labour, the state was
> obliged to maintain full employment and increase social welfare.[2]

The ramifications of this remarkable combination of factors in
terms of stop–go policies, balance-of-payments problems, sterling
crises and increased public spending are well-known, as is the fact that
by the early 1960s a much belated re-think was taking place.
Chastened by the Suez débâcle, the extent of overseas military commitments was now no longer beyond question; financially and politically,
reorientation towards Europe was under active consideration, with
inevitable implications for the special relationship with the United
States, for the importance of sterling and for residual tendencies
towards imperial posturing on a global scale; a move towards indica-
tive planning and industrial reorganization was gaining strength
within both major political parties; and recognition of the need for a
refurbished approach to industrial relations and incomes policy was
growing. Thus, by the beginning of the decade which would see the
potential of the North Sea turned into a live issue for the first time,
there were some signs of a fairly drastic reappraisal taking place with
regard to Britain's endemic problem of an economy which had signally
failed to readjust to the changed circumstances of the post-war era.

In the context of the present discussion, the important point is that
although such reappraisal was certainly taking place at this time, its
realization in practice was substantially delayed by constraints of the
same order as those that were responsible for the escalating problem
itself. On the international plane, any strategy designed to break out of
the impasse remained subject to constraints stemming from US
hegemony over the non-communist bloc's trade and financial system,

and not least from the fact that any unilateral move to rescue the British economy could have unpalatable implications for the dollar and for the stability of the international monetary system of which it was principal underwriter. In similar vein, supranational bodies like the International Monetary Fund (IMF), which had come to play such a prominent role in maintaining a viable framework for international trade and finance in the post-war era, could exert pressures for adjustments to domestic policy, usually in a deflationary direction, as the price of their support. Moreover, as Susan Strange has shown in her detailed study of sterling in the sixties, despite 'really inexorable processes' associated with the disintegration of empire and the rise to economic and political pre-eminence of the United States, despite the growing impact of the currency's changing international status and the concomitant practice of sustaining its position by concessionary inducement, the greater part of the decade was marked by the persistence of what she calls the Commonwealth Myth and the Top Currency Syndrome – political ideas which obstructed the adjustment of 'monetary attitudes and habits'.[3]

Such obstacles were accentuated by other aspects of British capital's unique situation at this time. Although the underlying strength of the working class, for example, was not deployed in open confrontation, it did mean that, as Purdy observes, 'the whole process of readjustment was particularly intractable and fraught with political risks, which dictated a cautious, piecemeal approach rather than a dramatic assault.'[4] More than anything else, however, the delay is attributable to the continued capacity of financial capital to secure the paramountcy of its own interest in the economic and political spheres. As Longstreth has argued, even after the return of a reformist Labour Government committed to industrial regeneration (in 1964), banking and the City retained a position of institutionalized dominance within the British state, a position that was exercised through the Bank of England and its connections with the Treasury and found expression in recurrent policy preoccupations with the balance of payments and the strength of the pound.[5] While the City did ultimately warm to its emerging role as an international market-place dealing mainly in non-sterling assets, in the meantime it solicited and received protection until 'the financial sector realized that it could survive devaluation';[6] throughout most of the 1960s, Longstreth maintains, the peculiar 'illogic' of the British state was displayed in the power of a 'political-economic fraction which has outlived its "world-historical role" by some decades'.[7]

Once again, the results of all this are well enough known – devaluation delayed until 1967; a National Plan 'sacrificed to the immediate exigencies of the balance of payments';[8] entry into the European Economic Commmunity (EEC) postponed until the early 1970s; and little progress with the task of industrial reorganization. State planning bodies, often 'located outside the institutional complex of central government', proved no match for the 'central axis' of Bank and Treasury which were thus able to retain and even enhance their control over short-term economic policy,[9] while foreign indebtedness and the 'multilateral surveillance' which accompanies it continued their dramatic growth.[10] With recurring exchange crises in 1960–1, 1964–6 and, of course, 1967, with the continued instability and chronic tendency to weakness of the balance of payments, the mid-sixties must indeed have seemed a timely moment for the realization that petroleum in one recoverable form or another might be lying beneath the sea bed of the Continental Shelf. As we shall see, moreover, such a possibility was also a timely one for other and quite specific reasons related to the position which oil was coming to hold in the British economy at this time.

For most industrialized nations, the process of post-war reconstruction had become increasingly dependent on oil during the 1950s, usually at the expense of other indigenous sources of energy.[11] Japan, destined to become the world's biggest importer, had rapidly discovered that the restored capacity to produce coal and hydroelectric power was not going to be sufficient to fuel recovery and by 1950 was adopting policies to ensure increased supplies of oil which, within a decade, came to rival coal in its economic significance. In Western Europe, too, a similar pattern occurred. Formerly dependent on coal, most countries turned increasingly to oil, initially in the form of imported and already refined petroleum products, but subsequently in the more flexible form of crude, as refining capacity grew towards self-sufficiency in the mid-1950s. Despite a less devastated coal industry, not to mention government attempts to afford it some measure of protection, Britain also moved increasingly towards becoming a heavily oil-dependent economy. In 1950, net imports of refined products and crude had totalled some 18 million tons and coal was still accounting for 90 per cent of primary energy requirements; thereafter, oil imports increased, albeit more slowly than in other Western European countries, until they reached around 50 million tons in 1960,[12] and Britain was on the road to the point where, some six years later, the market would be roughly divided between coal at 60 per cent and oil at 40 per cent.[13]

This penetration of oil into the economies of the industrialized nations of the West has been described by some observers as part of a conscious strategy designed to secure American hegemony over the political economy of the West in the aftermath of the war.[14] After all, five of the seven major oil companies are American, and through the Marshall Plan and even more direct control in Japan, the US authorities did have a considerable amount of leverage over the energy policies of the non-communist bloc. Such broader considerations notwithstanding, however, it is certainly the case that the companies themselves needed little encouragement. With her regulated and relatively high-cost indigenous production, the US had become a net importer of oil for the first time in 1948, with the result that the demand from Europe and elsewhere now had to be met from other and much lower cost sources, such as Venezuela and, increasingly, the developing oil fields of the Middle East. But pricing arrangements in the immediate post-war period remained much as they had been in the era when the United States herself had been the main supplier of a less oil-thirsty pre-war world, the delivered price being based on the expensive Texas rate and fictional costs of transportation from the Gulf of Mexico, regardless of the substantially lower production and delivery costs actually incurred elsewhere. As a result, the task of satisfying Europe's growing demand for oil offered the prospect of massive profits.

Under pressure from disgruntled European Governments and from officials of the Marshall Plan, which was having to foot the bill for oil imported at artificially inflated prices, the companies were persuaded to establish separate 'posted prices' for oil emanating from Venezuela and the Middle East in 1950. As Odell stresses, however, even then the new pricing system left the supply price well above that at which the companies would need to sell in order to receive an adequate return on investment, with the result that the market in oil remained very attractive.[15] And not just for the huge multi-nationals: as the fifties wore on, the so-called 'independents', which had previously restricted their operations to the United States itself, began to break into the low-cost overseas production areas, with the encouragement of host Governments anxious to break the stranglehold of the Seven Sisters,[16] and with the initial objective of enhancing their profits at home by importing cheap supplies.[17] However, in 1959 the United States Government imposed statutory import controls, and the independents were therefore increasingly attracted to the European market, where they began to undercut the majors. Coupled with the Soviet Union's return to the world market towards the end of the decade and the emergence of new, prolific sources of supply like Libya in the early 1960s, these developments produced a glut in world oil supplies.

The results were to be profound indeed. For one thing, the situation of over-supply ushered in an era of cheapening oil, which both ensured growing European dependence on this energy source during the 1960s (by 1965 oil imports to Western Europe were already accounting for around 33 per cent of primary energy consumption,[18] and by the following year it had overtaken coal in significance)[19] and gave Europe a competitive edge over the United States in terms of energy costs, an advantage that would play no small part in subsequent developments during the 1970s.[20] No less important, the competitive pressure on the majors forced them to reduce market prices and then, in an effort to compensate for the profit erosion thereby created, to cut the 'posted prices' at the Persian Gulf, the prices upon which the share of profits accruing to the producing states was calculated. The most immediate consequence of this step was the formation of OPEC in 1960 as an attempt to resist further reductions. Less immediately, declining profitability and these first signs that an inter-state cartel might now emerge to match the one formerly operated so successfully by the multinationals, sent the latter in search of possibilities for diversification and for less politically fraught settings in which to pursue their activities. By the mid-sixties, for example, Exxon had embarked upon a major exploration programme outwith the OPEC region, and although there is considerable dispute as to the nature of the prescience which brought the companies to the North Sea at a time when production costs would almost certainly exceed market prices, it was at this point that interest in the British Continental Shelf began to emerge, particularly in the wake of a major gas discovery in Holland's Groningen province during 1959.[21]

Against even this sketchy background of complex interlocking forces, it becomes possible to make some sense of the way British policy towards the North Sea began to develop from the early sixties onwards. Most obviously, of course, the formation of OPEC and the Suez demonstration of declining military influence over the world's major oil-producing area meant that the issue of security of supply now took on new importance, even if the real significance of the threat in this context was not fully appreciated until much later. Much more crucial, however, was the fact that although oil was indeed becoming cheaper, the increased reliance generated by this changed market situation and by oil's attractiveness for a variety of end uses meant that the issue became inextricably bound up with the balance of payments, to which oil imports were contributing a net deficit of around £300 million by 1965.[22] Nor, paradoxically, was this aspect of the matter helped any by Britain's own involvement in the international oil business through Shell and BP, an involvement which spawned a perceived interest in maintaining world prices and, by dint of the multinationals' dominant role in the British market, one which meant that Britain was paying around 25 per cent more than her European competitors for the oil which she imported.[23] Yet another twist was added by the assumption that, whatever the North Sea might yield, Britain would remain a major oil importer for a long time to come. The dubious implication was that the imposition of onerous terms on the companies now about to explore the UK's Continental Shelf might incite OPEC to follow suit, to the detriment of British overseas oil interests and, inevitably, that of the balance of payments.[24]

In the light of all these considerations, it is not perhaps surprising that, right from the outset, British North Sea policy fell prey to what Petter Nore has called 'the fatal historical consequences of the British obsession with the balance of payments and the defence of the pound'.[25] As we have already seen, the nexus between financial capital and strategically located agencies within the state bureaucracy kept these totems to the forefront throughout the greater part of the decade and, apart from delaying devaluation, perpetuating stop–go policies and generally impeding the restructuring of British industry, it also inevitably affected the authorities' approach to the possibility of a new source of wealth lying beneath the sea bed. Thus, for example, Adrian Hamilton points out that the inter-departmental committee (Ministry of Power, Treasury and Cabinet Office) which preceded the first round of licensing in 1964 opted firmly for as rapid exploration and exploitation as possible in order to meet the 'overriding need' for a balance of payments. Indeed, he goes on to suggest, so great was the Treasury's preoccupation with this side of things that it even positively favoured foreign investment during the early years 'because of the immediate benefits to the capital account'.[26] Similarly, when the Department of Trade and Industry was called upon, almost a decade later, to justify the early policies adopted by its predecessor, the Ministry of Power, the balance of payments ranked alongside security of supply in its account of the reasons for the basic decision 'to encourage the most rapid and thorough exploration and economical development of any petroleum resources on the UK shelf':[27]

The United Kingdom would gain substantially from the production

of indigenous oil or gas, providing an additional and secure source of
primary energy and benefiting our balance of payments. Retained oil
imports were then costing the United Kingdom about £300m. a year
in foreign exchange.[28]

In this way then, speed became one of the dominant considerations
in the formulation of the policy to be followed in the early rounds of
North Sea licensing, the first in 1964 under the auspices of a dying
Conservative Government, and the second under its Labour successor
in the following year. One reason for adopting a discretionary licensing system as opposed to a straightforward competitive one, for example, was that it was thought to give Governments a greater degree of
leverage in securing their objective of rapid progress, since applicants'
plans in this respect could be scrutinized in advance. Similarly, low
annual rentals, a requirement that half of each licensed area must be
surrendered after six years and the inclusion of some of the more
speculative independents in the allocations can be seen as motivated, at
least in part, by the desire for speed. Nor was any secret made of the
priority to be given to this goal. The licensing system, outlined to the
Commons by Conservative spokesmen and followed by Labour with
some modifications on matters such as public participation, was
designed 'with a view to securing the efficient and prompt exploitation
of whatever resources may be below the sea bed',[29] while encouragement of efforts in that direction was to be the first criterion in the
granting of concessions.[30]

In the event, some measure of the speed at which the enterprise took
off can be gathered from the fact that by the end of 1965, ninety
licences covering 475 blocks and some 42,000 square miles of the UK
sector had been granted – for an initial outlay, it may be added, of a
little over £2.5 million on the part of the industry.[31] And rapid success
there certainly was too. In late 1965, BP made an important find of gas
in what became known as the West Sole field, and this was soon
followed by an even bigger discovery by Shell/Esso at Leman Bank.
Indefatigable and Hewett were added to the list in the second half of
1966, and by the end of the next year it was possible for a White Paper
on fuel policy to predict that, by 1975, offshore production of natural
gas might be running at 4000 million cubic feet per day, nearly four
times the then current rate of gas consumption and around 15 per cent
of the total demand for energy.[32]

These discoveries did nothing to diminish the sense of official urgency surrounding North Sea exploration, nor to cause second

thoughts about how expeditiously to proceed with respect to what had already been discovered. Making use of its effective monopoly powers over gas purchase, the Labour Government negotiated a 'cost-plus' rather than a market price for North Sea gas in the late 1960s, but in so doing it also endorsed rapid development, both as a sweetener for the industry, which has its own economic reasons for favouring speed, and on the grounds of a by now familiar official logic surrounding balance of payments, benefit to the economy and the maintenance of incentives to further exploration. As Robinson and Morgan comment with regard to the *quid pro quo* aspect of the affair. 'Given that "low price" was evidently regarded as a fixed point in the natural gas negotiations, it was necessary for some other variable to be adjusted if the producers were to retain an interest in the British offshore area. . . .' Indeed, they point out, such government interference as there was at this time took the form of attempts to persuade the Gas Council not to obstruct the companies' plans for rapid exploitation.[33] Not that this body lacked its own rationale for going along with speed. Backed by the Treasury, with its eye inevitably on balance-of-payments savings, the Gas Council was embarking upon a massive programme of expansion and conversion to natural gas within a projected timescale of ten years.[34] In more general terms, the Government's view on the advisability of building up supplies from the North Sea as rapidly as possible was stated quite baldly in the 1967 White Paper:

> At this stage a rapid build-up in supplies is envisaged, on a basis allowing for the absorption of all that the fields so far discovered are expected to produce in the mid-1970s. . . . Most of the natural gas available will go to the premium markets where it will largely be displacing oil, but there will be some supplies to bulk industrial users. . . .
> This policy will mean a shorter life for the gas fields than a policy of slow depletion and will involve using some of the gas in markets where the resource savings are relatively low. The Government believe that these disadvantages are outweighed by the value of giving an incentive to the further exploration needed to improve our knowledge of the ultimate reserves available, and by the benefits to the economy and the balance of payments which a fast build-up of supplies will bring.[35]

The same month that saw the publication of the White Paper also witnessed another more momentous event, the devaluation of sterling. At last, the $2.80 parity was abandoned, and along with this step came a review of public expenditure envisaging mainly a planned withdrawal from the Far East and the Persian Gulf and a publicly avowed reassessment of the pound's role as an international reserve currency. As Brittan explains, 'The illusions about Britain's financial and military role in the world – of which Mr Wilson had originally been the foremost exponent, and which had held the country back for so long – were at long last jettisoned.'[36] Even then, however, the respite was slow to make itself felt. A flight into gold (occasioned as much by the increasingly shaky position of the dollar as by the British decision), the diversification of countries in the Overseas Sterling Area into other currencies and, more than anything else perhaps, fears that further devaluation of sterling was yet to come all combined to maintain pressure on the pound and the balance of payments in the wake of devaluation.[37] The price in terms of the wider connotations of Britain's debtor role was high as well, with the IMF at one point, in 1969, seeking to impose what some saw as 'banana republic' conditions as the price of further support.[38] Thus, as Jessop points out, 'Even after the inevitable but long-delayed devaluation of sterling in 1967, priority was still given to continued deflation in order to repay international debt and accumulate a payments surplus rather than resort to economic planning and industrual reorganization.'[39] Towards the end of the decade, moreover, the Government became increasingly impaled on the other horn of a peculiarly British dilemma, as workers successfully began to resist the squeeze on incomes resulting from devaluation and the deflationary policies which followed it. The resulting spiral of wage–price inflation was to assume more and more importance in the years that followed.

It was against this somewhat unnerving background that, in September 1969, Labour announced what would turn out to be the last round of licensing predicated on the assumption that the petroleum resources at stake consisted solely of natural gas. Having already carried out a policy review, officials had reached the conclusion that no further large fields were likely to be found in the Southern Basin, and since no assessment of the Northern Basin's potential was yet available, it was therefore decided that the emphasis should be on encouraging the industry to develop smaller, higher-cost gas fields. The result was a relatively restricted round of licensing, in which the terms, apart from some further steps to promote state participation and an increase in payments to allow for inflation, remained unaltered, as did the method of allocation by discretion, 'since the basic policy was still to encourage the rapid and thorough exploration and development of the United Kingdom shelf'.[40] Within a month of the round's completion, however, the rather pessimistic premise on which it had been based was shattered in an unexpected way when Phillips announced that appraisal work on an oil strike in the Norwegian Ekofisk field had revealed a production potential of 7 million tons per year. From then on, oil was to hold the centre of the stage in the dramatic events of the 1970s.

The Seventies and the Oil Rush

Under the Tory administration which came to power in June 1970, events on all fronts seem to have converged almost inexorably both to accentuate the British crisis and to sustain the impetus towards rapid offshore progress. Indeed, it was under the auspices of this Government that developments on the international and national planes conjoined to produce the so-called 'energy crisis' which, for a time, appeared to encapsulate some of the most basic exogenous and domestic issues underlying the predicament of British political and economic life – the confrontation with the miners in late 1973, combined with quadrupling of oil prices by OPEC and the spectre of a cutback in oil supplies to the West. Even though this combination of developments brought down a Government, however, it was only the culmination of processes which had been in train for some time, and it is with these processes that analysis of the 1970s oil rush in the North Sea has to begin.

On the international front, the first point of major importance is that, by 1970, the long post-war boom enjoyed by the capitalist economies of the world was at an end, while, in addition, the international trade cycle had entered upon a new era of heightened synchronization in which recession in one country was no longer to be as easily offset by boom or relative escape in another.[41] Faced with the ogre of inflation, many Governments in the developed world responded with fairly harsh demand-management policies, which produced a significant general recession in 1970–1, followed by a hectic inflationary boom in 1972–3 as monetary-based expansion took over. The boom, however, was short-lived. Even before OPEC, the customary whipping-boy in this context, jacked up the price of oil in late 1973, there were signs of a further recession on the way as a result of anti-inflationary policies being adopted in the US, Germany and Japan. By 1974, the slump was acute, and even though there was sufficient recovery to stimulate a period of fairly rapid growth in

1976–77, the predominant pattern up to the end of the decade was to be one of only feeble recovery.

Connected with, and indeed contributing to, these broad developments was another factor, namely, the collapse during the early 1970s of the Bretton Woods system of international payments which had played such an important role in the stable internationalization of post-war trade under the dominance of the dollar. The latter's hegemony, Skidelsky points out, however, had never been as securely based on a self-sustaining surplus as the pound had been in its nineteenth-century heyday, and from the late 1960s onwards it came under growing pressure.[42] Increased overseas expenditure (particularly as a result of the Vietnam war), combined with the growing competitiveness of the German, Japanese and other economies, produced a new pattern of deficit in the American balance of payments, while old European rivalry reasserted itself in the form of a French challenge aimed at the dollar's convertibility into gold.[43] In 1971, a speculative run on the US currency led to its devaluation and, at the end of the year, to a new agreement on exchange parities, the Smithsonian Agreement. However, renewed speculation maintained the pressure, and over the next eighteen months the system progressively collapsed as more and more countries opted for floating exchange rates. (Britain took this step in June 1972.) Thus one of the central planks in the platform holding up a stable framework of international trade and finance was now in a fair state of disintegration.

All of the above factors were to play a part in yet another external development of significance, the assertion of strength by OPEC, which, throughout the 1960s, had enjoyed only the limited success of preventing further price reductions and of imposing an ineffective boycott on the West during the Six Day War of 1967. Thus, for example, dollar devaluations provided a cogent case for an increase in the 'posted price' of oil, since it was to that currency that the price was fixed. Similarly, the emerging pattern of inflation in the West meant that the price of commodity imports by the oil-producing countries was liable to increase faster than the oil prices agreed between them and the companies, thus buttressing the case for an upward trend still further. Paradoxically, too, the boom of 1972–3 served to keep the demand for oil high and hence to lend credibility to the spectre of imminent shortage which could only be exacerbated by threats of cutbacks or even embargo. Not least, as several authors have observed, the growing economic predicament of the United States led Washington to be sympathetic, at the very least, to price rises that would reduce the competitive edge in energy costs enjoyed by industrial rivals for so long. The point is put less ambiguously by Odell:

> given the fact that the USA was fed up with a situation in which the rest of the industrialized world had access to cheap energy . . . it deliberately initiated a foreign policy which aimed at getting oil-producing nations' revenues moving strongly up by talking incessantly to the producers about their low oil prices and by showing them the favourable impact of much higher prices. It was, of course, assured of the co-operation of the largely American oil companies in ensuring that these cost increases, plus further increases designed to ensure higher profit levels for the companies, were passed on to the European and Japanese energy consumers, so eliminating their energy-cost advantage over their competitors in the United States.[44]

The complex chronicle of events surrounding the dramatic rise in oil prices during the early 1970s has been well documented and described by Anthony Sampson, Joe Stork and others.[45] Ironically, perhaps, it was Libya, one of the 'completely new sources' which had allowed the 1967 White Paper on fuel policy to adopt an almost self-congratulatory tone about the extent of Britain's diversification away from reliance on the Middle East,[46] that started the process off, when the newly installed Colonel Gaddafi imposed price rises during the second half of 1970. Thereafter, a leapfrogging effect between the Mediterranean and the Persian Gulf set in, until the Teheran and Tripoli agreements of 1971, which provided for increases bringing the price of Gulf and Libyan crudes up to $3.00 and $3.30 per barrel respectively. These agreements held for approximately two years (instead of the projected four), but the issue of price became more and more conflated with other issues which were rapidly destabilizing the world oil economy. OPEC members, for example, now appreciated that an enhanced share of the rent stemming from their oil resources could best be captured by increased state participation or nationalization, with the result that a wave of negotiations on this score came to overlay the increasingly fraught question of price. Use of oil as a political weapon against Israel also compounded the issue, culminating in the announcement of a progressive embargo by the Arab members of OPEC during the October war of 1973. By then, the price agreements worked out two years earlier were already teetering as a result of continued demand, envisaged shortages and the rapidly escalating sums being offered on the 'free' market, so that the jump in OPEC prices to $5.12 per barrel in October 1973 cannot have come completely as a surprise. The worst, however, was still to come. Two months later, the Gulf members of OPEC met in Teheran and, despite opposition from the Saudis, raised their price to $11.65 per barrel.

As Petter Nore points out, the significance of the events of 1973–4 lay not so much in their immediate impact, temporarily panic-inducing as they may have been, as in the fact that they comprised 'the first major decision about the development of the world's economic and political system which for at least two centuries was *not* taken either in Europe or in North America'.[47] For Britain, however, their coincidence with domestic developments which were rapidly coming to a head at the same time was to prove particularly momentous, even in the short term. As we shall see, moreover, even before the crunch came in late 1973, the combined effect of such external factors and the continuing internal problems of the British economic and political systems was to maintain the pressure for speedy development of the resources of the Continental Shelf.

The record of the Conservatives' term in office from mid-1970 to the beginning of 1974 is indeed a sorry tale from almost any point of view. Despite some success in facilitating the redirection of British capital (notably by entry to the EEC) and in enforcing a measure of rationalization through initially restrictive policies backed by legal constraints on trades unions, the overall record in terms of reversing or even halting the decline was pretty minimal. According to one useful summary of the period in question, the rate of profit fell from 8.7 per cent to 7.2 per cent between 1970 and 1973, while British exports of manufactures dropped from 10.8 per cent to 9.4 per cent of such world trade, and the rate of inflation rose from 6.3 per cent to 9.1 per cent.[48] Whether because of failure to control the latter factor or because of efforts to defend the new pattern of exchange rates agreed in 1971, there was a renewed bout of speculation against the pound in the following year, with the result that the exchange rate was finally floated at precisely the point when the world context in which it would do so was about to undergo dramatic change.[49] Concomitantly, the substantial surpluses on the balance of payments which the Government enjoyed during its first two years in power, arguably as a consequence of the delayed effects of devaluation, were already disappearing in 1972, and by the end of the next year, the current balance was showing a sizeable deficit of around £1 billion.[50]

Nor was the situation in general helped very much by the celebrated U-turn executed in mid-term. For one thing, the speculative boom which ensued served to pull in imports more rapidly than it boosted exports, and this at a time when the increasingly synchronized cycle of international trade was witnessing an explosion in commodity prices. No less important, as Glyn and Harrison emphasize, increased investment by both private capital and Government did not, in the main, find its way into the manufacturing sector where it was most needed,[51] while the state's activities in this context led to a substantial growth of the public-sector borrowing requirement (PSBR) from around £1.4 billion in 1971 to over £4 billion in 1973.[52] Most of all, a change of tactics (if not of heart) with regard to the issue of wage regulation failed to head off the growing confrontation between Government and strategically placed groups of workers, what has come to be known as 'the defensive strength of the working class' finally leading Edward Heath to call and to lose a general election in February 1974.[53]

The history of North Sea oil is interwoven with the story of these years in several different ways. At the most obvious level, of course, the link is evident in the effect which the upward movement in price had upon the attractiveness to the industry of relatively high-cost areas like the North Sea and Alaska. Already drawn in such directions by the greater security of supply which, among other things, they might offer, the oil companies had been caught between this consideration and declining rates of profitability for some time, with the result that only increased prices could adjust the balance in the optimal direction. Indeed, according to some observers, their concern on this score was one of the reasons for what is seen as the industry's collusion with the price-raising process in the Middle East.[54] Whatever their role in that side of the matter, however, there is no doubt that from 1970 onwards, the first really significant discoveries of oil started to be made in the Northern Basin of the British sector of the North Sea. Although Amoco had struck the Montrose field in September 1969, it was not until just over a year later that the really substantial prospects began to open up with BP's discovery of the Forties field. In February 1971, Shell/Esso found a small field (Auk), and in mid-summer the same consortium made a major find, subsequently to be known as Brent. With the addition of the smaller Argyll field, located by Hamilton Brothers and notable for being the first field in production (in June 1975), by the end of 1971 it was becoming clear that the UK had fairly sizeable oil reserves on its maritime doorstep. Significantly, too, this was at a time when, despite an earlier forecast of a 41 per cent share of the British energy market going to oil by 1975, the actual share already captured was nearer 50 per cent.[55]

At this juncture the different strands of the story were drawn even more closely together by the Government's announcement of a fourth and major round of licensing to be held in late 1971 and early 1972. Prompted by what was thought to be a fall-off in exploration activity during 1970–1 (despite the finds made thus far), this round once again gave priority to speed and revealed how closely the question of the North Sea's resources and the pace of their exploitation was tied up with the wider national and international developments already described. Thus, for example, now that oil had been found, the attraction of the 12½ per cent royalty to be collected on a rising market oil value as opposed to the cheaper cost-related price of gas did not escape the attentions of a Government already caught in a spiral of growing public expenditure. The Department of Trade and Industry later explained the official thinking of the time thus: 'The prospect for the Exchequer in the shape of receipts from royalties would be enhanced if rapid and thorough exploration led to more and earlier discoveries.'[56] Equally, the developments within OPEC inevitably penetrated the issue in predictable fashion. In terms redolent of earlier days, a decision not to harden licence conditions was justified on the grounds that British oil companies were now embroiled in delicate negotiations with the OPEC producers, and 'ministers were inclined still to take a careful view about not harming our own oil interests abroad.'[57] OPEC actions and their implications for the abiding constraints of the balance of payments were also called upon in support of the case for speed:

> At the same time, following the success of the Libyan Government in September 1970 in increasing its 'take' from oil company concessionaires, other members of OPEC succeeded in following suit. The addition to the foreign exchange cost of the United Kingdom's oil imports was estimated to be about £200m. and on balance of payments grounds alone there was every incentive to hasten the search for indigenous supplies.[58]

This licensing round was subsequently to become part of a near scandal when the Public Accounts Committee of the House of Commons produced a report on North Sea oil and gas in early 1973.[59] Therein it was revealed that although a preliminary experiment with auctioning some licences had pulled in £37 million for a mere fifteen blocks, a further 267 had then been allocated for the paltry sum of £2.7 million under the traditional discretionary system. Moreover, it was suggested, the whole policy up to and including this round had been badly flawed by its failure to take account of the combined effect of several factors on both the revenue and balance-of-payment implications of the North Sea. Transfer pricing within the vertically integrated structure of the major oil companies, OPEC's high 'posted prices' and UK rules on relief from double taxation had led to the accumulation of tax losses approaching £1,500 million on Middle East operations, losses which could be carried forward and set off against liability to the corporation tax payable on future North Sea profits. Some of a UK group's capital investment outwith the area could be similarly offset, while avoidance of corporation tax would mean a large remission of profits overseas by foreign companies (predicted to be responsible for two-thirds of North Sea production by 1980), to the detriment rather than the benefit of the balance of payments. Thus the Committee estimated that little tax revenue could be expected from profits even by the next decade, and that the only 'certain' revenue would come from the low royalty and even lower licensing payments. All in all, it concluded: 'Under the present arrangements the UK will not obtain either for the Exchequer or the balance of payments anything like the share of the 'take' of oil companies on the Continental Shelf that other countries are obtaining for oil within their territories.'[60]

Even before it was overtaken by events, the Heath Government was moving towards closing some of the major loopholes revealed by this Committee.[61] In the meantime, however, in the wake of the 1971–2 round of licensing, progress in the North Sea itself had been rapid. During 1972, the Beryl and South Cormorant oil fields were added to the list, as was the British share of the Frigg gas field, while the following year saw the addition of such now familiar names as Thistle, Heather, Dunlin and Piper, all oil fields. The less well-known Maureen, Hutton and Alwyn fields were also found in 1973, and during the Government's last full month in office (January 1974), a consortium dominated by Burmah, Chevron, and ICI Petroleum made a substantial discovery at Ninian. By then, indigenously produced natural gas was already meeting 95 per cent of the country's gas requirement and around 12 per cent of the total inland energy demand,[62] while estimates published some months earlier had tentatively canvassed an oil-production level of 70–100 million tons per year by 1980, a figure that would approximate to Britain's total oil requirement.[63]

Whatever the benefits to be derived from such developments – and it must be remembered that at this point actual production of oil had not yet started – the Conservatives were not to comprise the Government that could pin its hopes to them, at least in the short-term, for other events already alluded to were now moving rapidly towards their climax. Against a background of expected shortages and spiralling inflation in the West, OPEC had increased its prices by some 70 per cent in October 1973, while at the same time, the organization's Arab members had announced the beginning of their progressive embargo. In November, the overall cutback was standardized at 25 per cent,[64] while closer to home, the British miners turned down the National Coal Board's latest pay offer and imposed a ban on overtime. The Government responded by declaring a state of emergency and imposing a three-day week on manufacturing industry from the end of December. By then, however, OPEC's prices had more than doubled once again, and with the miners' resolution holding out into the new year, Britain's political and economic crises became fused with the wider international 'energy crisis'. As more than one commentator pointed out at the time, had it not been for the apparent unreliability of high-priced oil supplies, the shortfall in coal production could have been contained with a less drastic effect upon the economy.[65] Nor, we may add, is it certain that, on its own, the confrontation over pay would have created the appearance of a crisis sufficiently grave to tempt the Government into an election on the issue of 'Who runs the country?' Conversely, had it not been for the capacity of strategic sectors of the workforce to put up effective resistance against the squeeze on their standard of living at this time, developments on the international plane would not, in all probability, have had such a sharp effect upon domestic politics. In the event, however, both sets of factors did combine, and the Tories lost the election, bequeathing to their successors a situation in which, as we shall see, maintaining full speed ahead with the exploration and exploitation of the Continental Shelf would continue to hold its attraction.

The world background against which Labour took over the governance of Britain from 1974 to 1979 was not propitious, to say the least. Already firmly in train before OPEC's actions, the first really major and generalized recession since the Second World War was beginning to bite hard on the capitalist economies of the world by the spring and summer of 1974. With inflation spiralling towards an average of 15 per cent,[66] most countries hesitated over responding with anti-cyclical measures to restimulate their economies, and indeed, with some minor exceptions such as Canada and Sweden, they resorted to deflationary domestic policies designed to offset balance-of-payments deficits and to reduce the rate of inflation.[67] The result was a collapse on a scale unparalleled since 1929–32. According to Mandel, unemployment in the capitalist countries rose towards its (for then) peak of 17 million in 1975–6, while in 1975 the volume of their exports actually fell for the first time since the beginning of the long period of post-war expansion.[68] Industrial production fell by 10 per cent between the middle of 1974 and the autumn of 1975.[69]

When the recovery began in late 1975, it was to be slow, very hesitant and certainly insufficient to recreate the general optimism that had pervaded most of the 1960s. Fears of refuelling inflation now made Governments reluctant to adopt expansionary policies on the scale necessary to engender major restimulation and to eliminate unemployment. Indeed, the recession of 1974–5 marks the breaking point in the general post-war commitment to relatively full employment,[70] a commitment that had played no small part in enhancing the strength of the working class in the industrialized societies of the West. Equally, increased recourse to monetary targets as a reaction to the apparent failure of demand-management policies meant that the large budget deficits accumulated in even the hesitant efforts to stimulate recovery became increasingly unacceptable. Coupled with a lower than expected up-turn in productive investment, a return to protectionist policies and fears that some Third World countries might default on the large loans which they had taken out in the form of recycled petrodollars to finance their trade deficits, these factors combined to ensure that the recovery would not be the forerunner of a new and protracted phase of accelerated growth. Indeed, by 1977–8 the recovery was already faltering, and the stage was set for a further and possibly even more catastrophic recession.

Against this background it is not perhaps surprising that the record of the 1974–9 Labour administration in Britain should have been one that was dogged with difficulties. To be sure, the new Government came to power with brave enough plans for halting economic decline and for improving the country's competitiveness *vis-à-vis* its major industrial rivals, not to mention its plans for increasing social equality, but it very quickly ran into familiar short-term constraints stemming from international developments and from a constellation of factors specific to the British situation. On the wider front, the first problem was, of course, the inheritance of an escalating balance-of-payments deficit which rocketed to nearly £3.5 billion on current account in 1974, an oil deficit of over £3 billion making a substantial contribution to a gap on visible trade which had now stretched to around £5.5 billion.[71] Moreover, while the deficit was easily enough financed, not least because of the inflow of short-term funds from the OPEC surpluses to London, other less auspicious trends were also involved. For one thing, the scale of Britain's short- and medium-term foreign indebtedness was growing rapidly, having jumped from around $3 billion to just over $7 billion between 1973 and 1974.[72] No less important, just as short-term inflows to the sterling balances could readily reverse themselves, particularly after the termination of dollar-value guarantees in 1974, so too the City's new-found role as a major financial centre dealing in Eurocurrency and other non-sterling currencies meant that deposits could be switched around with considerable facility. As a result, both sterling and the reserves were vulnerable to the highly volatile and often speculative nature of short-term capital movements.[73]

On the domestic plane, the most immediate constraint on the new Government emanated from the circumstances of its own election, namely, a backlash against the restrictive and divisive policies of its predecessor. In consequence, early recourse to the severe contractionary policies being adopted elsewhere was not a viable political alternative, at least for the time being, and public spending continued to grow 'despite the increase in public-sector borrowing which this entailed, and despite the acceleration of inflation, which it was clear could only be abated by following the example of other countries in administering sharp deflationary pressures to the labour market'.[74] Thus, while the full force of the world recession was delayed in the British case, the temporary reprieve was bought at a high price. The wage–price spiral continued on its upward trajectory, and inflation broke through the double-digit barrier to reach around 16 per cent in 1974 and over 22 per cent in the following year. At the same time, with the Government's initially neutral stance on overall expenditure, the PSBR leapt up from £4 billion to £6.5 billion in 1974, and in the following year grew even further to some £10.5 billion. As one commentator subsequently observed, albeit without the advantage of knowing what would happen at the beginning of the following decade, at this point all the main economic indicators were registering record 'worsts', and 'the long decline of British capitalism had reached its nadir.'[75]

The rest of the story will be familiar enough to anyone who recalls the fraught years between 1975 and 1979. Attempts to stem the tide began in the former year with public expenditure cuts and a sterling crisis, which was rapidly followed by a voluntary agreement on wage increases. As Paul Ormerod points out, within months of Labour's being re-elected in the second election of 1974, 'the Government's economic strategy was being constructed with reference to the politically motivated and misguided economics of the City and the financial markets. . . .'[76] A further and dramatic run on the currency in 1976 increased the pressure still more, with resort to massive IMF support bringing stringent monetary target conditions in its train. Thereafter, the record was one of policies heavily dominated by the Treasury, of substantially attenuated efforts to implement a planned strategy for industrial regeneration and of persistently high unemployment. Although inflation was temporarily brought under some semblance of control during the latter part of the Government's stay in office, this was achieved largely at the expense of cuts in real living standards, cuts which finally provoked the famed 'winter of discontent'. Equally, while exports and the balance of payments situation did show a marked improvement from 1977 onwards, this did not represent any dramatic turn-around in underlying competitiveness. Whereas world trade in manufactures increased by 30 per cent between 1975 and 1979, the UK share remained virtually static at around 9.4 per cent. *Mutatis mutandis,* penetration by imports of manufactured goods to Britain rose from 22 per cent to 26 per cent of home demand during the same period, while their volume increased by over 50 per cent.[77]

Amid the fairly unremitting gloom of these years, oil from the North Sea seemed to offer one of the very few rays of sunshine. Indeed, in what has been described as a particularly 'unctuous coupling of God with the interests of the British people', Prime Minister Callaghan at one point reportedly even alluded to the subject in terms which suggested that the Deity was once again on Britain's side.[78] Supernatural allegiances apart, however, it is not difficult to see why offshore oil should have provoked some mild euphoria at this time, for it seemed to offer at least partial solutions to national problems just at a point when they were becoming particularly acute. The underwriting, not to mention the repayment, of foreign loans, for example, could be greatly facilitated by possession of such increasingly valuable assets on the Continental Shelf, while the prospect of additional income to the coffers of the state held obvious attractions for a Government espousing, or being pushed into, more and more revenue/expenditure consciousness. As ever, too, a reduction in the level of oil imports would greatly benefit the balance of payments. With luck, North Sea oil might even play a leading role in restimulating the economy and regenerating Britain's industrial base. While it can be argued, as Guy Arnold puts it, 'that the political grasping at the North Sea oil lifeboat to save the economy is a major indictment of the performance of successive Governments over many years to solve anything at all',[79] there is no doubting the fact that in the particular circumstances of 1974–9, the mineral resources of the Continental Shelf were seen as an especially welcome sign of imminent rescue.

Even before the Government took office in 1974, it was apparent that despite plans for a fairly drastic tightening up of the offshore industry, a new Labour administration would not abandon speed as a policy priority. Speaking at the height of the 1973–4 'emergency', Mr Eric Varley – subsequently to become Secretary of State for Energy – expressed the view that the 'first objective' of the newly created Department of Energy must be 'the building up of alternatives to unreliable supplies of over-priced imported oil'.[80] Once in office, he and his colleagues stuck firmly to the target of rapid offshore progress, at least in the short term, and openly avowed this to be their objective. In one of the frequent tactical allusions to the fact that Britain's oil lay predominantly off the coast of Scotland, Dr Gavin Strang (Under-Secretary of State for Energy) told the Commons in March 1974, 'It is in the interests of everyone, including the people of Scotland, that we get the oil flowing in significant quantities as soon as possible.'[81] In July of the same year a White Paper on offshore policy announced, among other things, that measures to control physical production in the national interest would now be taken; but it at once hastened to explain that 'this does not affect . . . determination to build up production as quickly as possible over the next few years.'[82] Some six months later, the Secretary of State himself confirmed his intention of assuming, though not of using right away, powers to control depletion rates. In so doing, however, he again gave a series of assurances later officially described as 'intended to dispel uncertainty about depletion policy and to encourage a rapid build-up of self-sufficiency by 1980'.[83] 'It remains the Government's aim', he insisted, 'to ensure that oil production from the United Kingdom Continental Shelf builds up as quickly as possible over the next few years. . . .'[84] The target was to be the now more confidently predicted output of 100–140 million tons per annum by 1980,[85] and to achieve this the industry, ever sensitive to delay, was given promises which guaranteed that no delays would be imposed on the development of fields discovered before the end of 1975. No cutbacks would be made in production from such discoveries before 1982 at the earliest, and any finds made under existing licences after 1975 would not be subject to such cuts until 150 per cent of the field's capital investment had been recovered.[86] As the Department of Energy later observed, 'The effect of these assurances is that no delays can be imposed on the development of fields accounting for between a half and two-thirds of our estimated total reserves, and that no cutbacks can be made in production from such discoveries before 1982 at the earliest. . . .'[87]

In the meantime, then, the use of depletion controls to impose a major slow-down on North Sea development was not to be entertained. Since the cost of developing already discovered fields greatly exceeds the cost of looking for new ones, this decision must have come as something of a relief to the industry, which would otherwise have had to face what it regarded as unacceptably long time lags in recovering the heavier investment involved in the post-exploration phase. But this still left open the question of whether the rate of future exploration should be decelerated through the manipulation of licensing policy itself. Again, however, the Government could see good reasons for refraining from draconian measures in this context, and while it was decided to flatten out the fluctuations in offshore exploration activity by licensing smaller amounts of territory at more frequent intervals, two further licensing rounds were carried out, in 1976–7 and 1978–9 respectively. In all, licences covering a further eighty-six blocks were awarded, and official policy remained geared to rapid exploration.[88] Arguing that control through licensing policy would delay the acquisition of knowledge requisite to a sound depletion policy, the Government committed itself to the view that 'our objective must therefore be to secure that exploration continues as fast as reasonably practicable, provided the discovery of reserves is not allowed to lead, as a result of commercial pressures, to production earlier than the national interest requires.'[89]

Some writers on this subject have made much of the apparent change of heart which took place with regard to the pace of offshore activities under the auspices of this Labour administration,[90] and certainly it does seem that Mr Benn, Eric Varley's successor, did try to exercise his residual powers over the development of new fields.[91] From the point of view of the present work, however, the germane point is that although there were indeed firm intimations of second thoughts about the medium-term future, the emphasis on rapid progress remained central to both depletion and licensing policy in the meantime. Thus while Government moved towards much more effective control over the industry in other respects – principally by the introduction of a 45 per cent petroleum revenue tax, the erection of a tax 'ring fence' around the North Sea, and the formation of the British National Oil Corporation (BNOC) – speed continued to be of the essence right up to the end of the period with which this book is concerned.

And further success there certainly was, despite setbacks in meeting the scheduled dates of start-up for some fields, and in spite of a slackening in investment interest on the part of the industry.[92] By the end of 1978, a dozen oil fields were in production, with a total output of some 53 million tons of oil per annum. Equivalent to around 57 per cent of a total national oil requirement which had fallen back somewhat from its almost half-share of the energy market in 1974, North Sea oil production was by then running at a level approximating to 25 per cent of the country's total primary fuel consumption. By the end of 1979, it was predicted, these figures could reach 78 per cent and 35 per cent respectively.[93] On the pre-production side, a further eight fields had been discovered and put under development before the close of the Government's last full year in office, while some forty-four other 'significant' oil discoveries had been made.[94] Although none of the fields under development was in the giant class of Forties or Brent,[95] the overall picture was nonetheless one that augured well for net self-sufficiency by 1980.

The sought-for benefits from the North Sea, and particularly from its oil potential, also began to have a significant effect during Labour's stay in power.[96] The collection of royalties on oil began in 1975–6, and, with a yield of some £22 million from gas, produced a total royalty revenue of £66 million. Over the next three financial years, this overall total rose to £288 million. Because the structure of corporation tax and petroleum revenue tax postpones payment until capital and other allowances have been made up, a system 'designed to encourage development by allowing companies to recover their heavy capital costs early in the productive life of oil fields',[97] their contribution was rather slower to accumulate. But by 1978–9, they too were yielding well over £200 million per year. Most of all, however, it was during this period that the North Sea began to have a significant impact on the perennial problem of the balance of payments. In 1977, the value of oil exported and of imports saved came to over £2 billion and, for the first time, exceeded the costs of imported goods and services associated with the North Sea programme itself. Added to this, indigenous gas production was estimated to have been worth a further £2 billion at the then current price for imported oil equivalents, while a significant net inflow of funds on capital account to finance investment continued to make its contribution to reversing the UK's traditional tendency towards net capital exportation. (Between 1972 and 1978, the North

Sea programme accounted for 47 per cent of the country's capital inflow.)[98]

By this point, the North Sea was coming to be seen almost as a last chance to regenerate Britain's economic performance. In March 1978 the Government published a White Paper, *The Challenge of North Sea Oil*, which merits quotation at some length for the way in which it shows how thoroughly the question of the UK's offshore resources had now become imbricated with the long-term problems (and hopes) of the British economy:

> North Sea oil provides a unique opportunity for Britain to improve her economic performance, raise her living standards, move forward to full employment, and develop as a socially just society. It will also put her in a stronger position to discharge her international responsibilities, not least in relation to developing countries.
>
> For too long, Britain had a weak balance of payments and slow production growth. This weakness has many and complex origins, but the inadequacy and poor productivity of investment in a number of sectors of United Kingdom industry has been an important factor. The deep world recession of recent years has added greatly to our longer-term problems. Following the fivefold rise in oil prices, and the heavy additional cost of importing oil, Britain borrowed heavily abroad to finance the balance-of-payments deficit. In addition, we suffered dangerously rapid inflation. The recession has highlighted the longstanding uncompetitiveness of some of our major industries.
>
> As a result of this long period of decline, our living standards, which were among the highest in Western Europe a quarter of a century ago, are now among the lowest. Governments have tried repeatedly to alter these unfavourable trends. The present Government, despite a difficult international economic environment, has already made real progress in turning the situation round.
>
> In the past year the country's financial position has substantially improved. Sterling is stronger. We are no longer net borrowers from abroad: and we are beginning to repay debt. The rate of inflation has been dramatically reduced — thanks to the remarkable efforts of the whole British people and particularly to the co-operation of the trade union movement in a voluntary incomes policy. Unemployment is still unacceptably high in Britain as in other countries: and it is only the financial support given by the Government through selective employment measures to a number of firms and industries that has stopped it from being much higher.
>
> The next task is to get production moving and to reduce unemployment.
>
> North Sea oil gives us a great opportunity to achieve this. Sensibly used in pursuit of the policies laid out in this White Paper, it can

enable us to rebuild the industrial strength of the nation, to the benefit not only of this generation, but of our children and our grandchildren. To change and improve our national performance now and in the years immediately ahead, even with North Sea oil, will not be easy: but we shall never have a better chance.[99]

Geared to such brave long-term endeavours as these, the White Paper went on to spell out an ambitious programme for industrial investment, public expenditure, the development of alternative energy sources, cuts in taxation and a more dynamic regional policy. While eschewing the idea of an 'oil fund', as such, the Government nonetheless stated its intention of reporting annually to Parliament on the specific issue of progress in deploying the benefits of North Sea oil according to plan, the first to be submitted in the summer of 1979. Had the administration survived to do so, it would also have been able to report on belated plans to increase petroleum revenue tax from 45 per cent to 65 per cent and to reduce the capital uplift allowance from 75 per cent to 25 per cent. Equally, it could have catalogued continuing progress towards estimated oil savings of £5.5 billion on the balance of payments by 1980 and of no less than £8–9 billion in the longer term. Less optimistically, it might also have counselled a degree of caution in light of the fact that the pound, reacting to its new status as a petro-currency, had been staging a remarkable comeback since the end of 1977.

Cautious or ebullient, however, such a report was not to be submitted, for in May 1979 the Conservatives were returned to power, and the saga of the North Sea's development becomes part of a much more contemporary history of which readers will scarcely require much reminding. Within a year, another major world recession was to be in full swing, and Britain was to prove no exception. Manufacturing output fell by a full 14 per cent during 1980, more than in any year during the Great Depression,[100] while unemployment, already rising above the 1.5 million mark by May of that year, would reach 2 million by the autumn.[101] By then, manufacturing industry's gross fixed investment in plant and machinery was falling for the first time since early 1976, and the investment intentions elicited from its members by the Confederation of British Industry (CBI) were showing their most adverse balance since the height of the previous recession.[102] Under the combined impact of stringent, if unevenly applied, monetary policies and adverse terms for external trade, more and more firms were forced into closure, while still others became increasingly dependent on the selective largesse of the high street banks and, significantly perhaps, of the Bank of England.[103] All in all, at the time of writing (the beginning of 1981) it seemed that the British economy was heading for a crash from which it might recover only at the cost of a decimated industrial base. Nor was there any shortage of financial and industrial commentators who felt that, this time, the economy and manufacturing industry in particular was finally dipping towards the point of no return.

In such circumstances, and with oil now coming ashore at the rate of around 79 million tons per year (worth a cool £2,230 million in government revenue during 1979), it might be thought that, at long last, the policy of successive Governments with regard to offshore oil might finally pay off just when such a windfall was most needed. Paradoxically, however, and despite the fact that the oil undoubtedly cushioned the majority of wage earners against the full force of the slump, Britain now became hoist with her own petard, as oil, along with high interest rates, drove the pound far beyond the exchange level justified by the relative uncompetitiveness of the British economy. As a result, a country which had for so long been told that it must export manufactures or die now found itself being strangled by its own good fortune in having substantial reserves of one of the world's most valuable commodities available on its doorstep. Moreover, with each successive rise in the oil price – in June 1979, it rose to $20.5 per barrel, and around a year later the OPEC average price was around $30 per barrel – this particular aspect of the problem worsened, even if it also meant an increase in the value of Britain's own highly priced reserves. Thus, as time wore on, more and more economic analyses of the nation's plight came to acknowledge that North Sea oil now had a stranglehold on the neck of British manufacturing. 'How North Sea oil is killing our industry' was the title given to one such discussion,[104] while another referred to a 'financial doomsday machine' which was merely substituting North Sea oil for other forms of output – with horrifying prospects for unemployment.[105] For some, the latter now became another inevitable price that, along with de-industrialization, might have to be paid, the best chances for salvaging any of the North Sea's much vaunted benefits lying in settling for cheaper imports and in encouraging foreign investment.[106] 'Our oil and the people mountain' was the sanguine title given to a review of one such depressing outlook for British industry at the end of 1980.[107]

To others, however, it did not seem that the industrial towel should be quite so readily thrown into the ring. By early 1981 calls for the establishment of an 'oil fund' specifically oriented to investment in the economy's infrastructure were growing,[108] while the chairman of Shell was urging – among other, more predictable things – a policy that would dedicate a significant part of North Sea income to industrial restructuring.[109] From mid-1979 onwards, moreover, deepening doubts had been expressed about the wisdom of speeding North Sea developments, with some observers counselling the Government to take a leaf out of the OPEC book by slowing down the rate at which offshore oil was being taken out.[110] Despite the practical difficulties involved, not least because of the assurances previously given, it was argued that this might be one of the only effective means of stemming the rise of the pound and helping industry. In similar vein, the CBI was treated to some plain speaking from the British Leyland chief, Sir Michael Edwardes, in November 1980. Why, he wanted to know, were the oil revenues not being used to help industry to become more competitive rather than to accelerate industrial decay and to pay for the consequences of industrial failure? With characteristic verve, he concluded: 'If the Cabinet does not have the wit or imagination to do this, if they cannot cope with North Sea oil, then leave the bloody stuff in the ground.'[111]

Despite the adverse effect which offshore oil has arguably been having on industrial performance, and regardless of the forthright advice being proffered from many quarters on this score, the Conservatives have not thus far taken particularly decisive action to reduce the speed of North Sea developments. Indeed, although falling demand and rising output meant that net self-sufficiency was imminent by the middle of 1980, the intention to hold a further licensing round, subsequently described as 'expected to quicken the pace of oil and gas exploration in UK waters',[112] was announced in May of that year. To be sure, there had been some misgivings – disagreements about how to maintain high levels of industry confidence and drilling activity without losing ultimate control over production had delayed the announcement from December 1977[113] – but when it came, the package was an attractive one. Around ninety blocks were to be put on offer, and, for the first time, companies were to be given the chance of nominating up to twenty Northern Sector blocks that they would like to explore. At £5 million each, these would raise some £100 million, a figure that rose substantially when the number of blocks involved was increased to forty-two, partly because Ministers were anxious to offer as many blocks as possible under this round and partly perhaps, as the *Financial Times* laconically remarked, because 'the additional premium payments will be welcomed by the Treasury.'[114] Although the award of

Government-nominated blocks was delayed into the New Year, by the end of 1980 the industry was already applauding the extent of the round which was 'going towards the level we need'.[115]

On depletion policy, the Government's stance seems thus far to have been characterized also by a considerable amount of ambivalence. Early in 1980, the Energy Secretary gave warning that cuts in production might have to be imposed in order to remove a 'hump' which could emerge in mid-decade. Six months later, however, his confirmation of this possibility was being described as 'a retreat, though not a total one', as a result of which the policy was retained in principle without any commitment to specific production targets.[116] The Government, one commentator later wrote, 'is afraid of spelling out a clear policy for depletion', a reluctance that he attributed to a mixture of Treasury 'obstacles', political and economic doctrines averse to interference with market forces, and 'concern not to impede further oil exploration and development by breaking faith with the oil companies'.[117] Thus the Varley guidelines referred to above were substantially maintained, and the main area for manoeuvre on the depletion issue continued to be the possibility of imposing delays on the development of finds made after the end of 1975.

Overall, then, it is a matter of some conjecture whether the present Government will willingly or successfully decelerate the pace of offshore activity. Nor is it difficult to see why there should be a marked degree of hesitation in this respect. Increasingly, for example, the question of Britain's North Sea assets has become caught up in fraught negotiations with EEC partners, France and Germany in particular pressing hard for preferential prices and increased supplies.[118] Indeed, the Foreign Office has reportedly become concerned that Britain may not be using her potential productive capacity to optimal diplomatic advantage, either within the European Community or in terms of helping to stabilize world oil supplies.[119] Similarly, while continuing problems in the Gulf region have not brought a shortage, largely thanks to reduced demand, they have nonetheless created an atmosphere of psychological insecurity which places the ready availability of 'secure' supplies from areas such as the North Sea at a premium. Not least, the continuing upward spiral of prices and profits has enabled the Government to enhance its short-term tax prospects from the North Sea very considerably by raising petroleum revenue tax to 60 per cent in mid-1979 and then to 70 per cent in the Budget of March 1980, while a special 'windfall' tax of 20 per cent on the value of oil produced was announced in November of that year. By late

1980, one estimate of future revenues was predicting a 'take' of more than £17 billion in 1985.[120] A few months later, the Inland Revenue received the largest single tax payment in its history, when BP deposited over £854 million as its half-yearly payment of petroleum revenue tax on the Forties field.[121]

To a Government obsessed with such issues as the level of the PSBR, this prospect of growing revenue from rapid North Sea production is unlikely to be a gift horse that will be looked in the mouth, save perhaps in cases where even more immediate savings can be effected by cutting back on BNOC's investment plans.[122] According to the *Financial Times,* for example, one factor 'not talked about very loudly in Whitehall . . . lest the prospect of such a bonanza distract from the need for fiscal restraint' is that the spectacular rise in expected revenues over the next couple of years will give incomparably more room for manoeuvre in fiscal policy during the later stages of the Government's life.[123] More cynically, Anatole Kaletsky has pointed out that 'an open and informed debate on oil production would inevitably draw attention to the fact that fulfilment of the election promise about cutting income tax to 25p by 1984 depends entirely on oil revenues.'[124] In like vein, it has been noted that the extra tax revenue to be anticipated from North Sea oil will correspond almost exactly to the cost of keeping 3 million unemployed by the middle of the decade,[125] while those who already see the available funds being 'squandered' on keeping down the PSBR rather than being developed in 'constructive investment'[126] will derive little comfort from a recent Wood–McKenzie prediction that, by 1985, Government revenues will be just about equivalent to the present borrowing requirement.[127] While there is no way of assessing suggestions that future revenues are maybe being deliberately underestimated so that they can be 'updated at a suitable political date',[128] there is no dearth of lingering suspicion that the electoral implications of maintaining offshore momentum may not have escaped the Government:

> If Mrs Thatcher is short of a hopeful thought for the New Year, it might be that with North Sea oil enhancing the real living standards of those in work – as it has continued to do this year – she might win her re-election in spite of 3 million unemployed and manufacturing industry further down Sir Terence Beckett's plughole. With the oil revenues nearing their peak by 1983 it ought not to be too hard to arrange a substantial reduction in the taxes too.[129]

Debate will doubtless long continue as to the real impact which

North Sea oil has been having on British political and economic life during the current recession. On the present showing, however, it is difficult to avoid two conclusions. In the first place, it seems that the Conservative Government, like its predecessors, has fallen victim to its own short-term horizons, horizons which, as Robinson and Morgan pointed out some time ago, may 'result in *faster* exploitation because of the desire of politicians to obtain benefits now rather than in the future'.[130] Thus the combination of external and internal factors contributing to rapid offshore development by no means excludes the immediate exigencies of a political system that exacerbates the difficulties involved in 'technocratic' planning for the long-term needs of capital accumulation. Indeed, the whole history of Britain's offshore oil and gas may well turn out to be a paradigmatic example of how bureaucratic decision making tied to politically defined objectives tends to preclude effective overall planning.[131]

Governments and their policies, however, are not guided just by some opportunistic electoral self-interest. As we have seen at several points in this chapter, they are also constrained by the forces which gain pre-eminence within the machinery of the state itself, forces which initially shaped North Sea policy in accordance with the interests of finance capital and its anachronistic concern for the value of the pound and for sterling's international role. While these preoccupations and their associated shibboleths may indeed have vanished, however, this does not mean that policy with regard to the UK Continental Shelf has broken free from the fetters of financial interest altogether. With the advent of monetary targets under Labour and their 'fetishization' under the Tories, such interests have retained considerable influence in general,[132] while 'the Treasury–Bank axis is still dominant in the administrative sphere.'[133] More specifically, while there may certainly have been a trend towards integration and interdependence between the financial and industrial sectors of the economy, as argued, for example, by Scott,[134] this fusion seems to have taken a limited and particular form in the context of the North Sea. As Noreng observes, with regard to both Britain and Norway, there has been 'a split between a group composed of the financial establishment and the industries participating in oil development, and a group representing the bulk of manufacturing not related to oil'.[135] Thus it is not perhaps entirely fanciful to see the earlier comments of hard-pressed industrialists and others as evidencing a residual dislocation between financial and industrial capital, with the former's influence over policy being bolstered by the dominant role accorded to multinational oil corporations in the development of the Continental Shelf. If such a view is justified, the speed of offshore progress may well go down in history as another case of *plus ça change, plus c'est la même chose.*

Speed and Dependency

The final question to be addressed in this chapter is whether speed has affected the relationship between successive British Governments and an offshore oil industry dominated by multinational corporations. To be less provisional about it, I wish to suggest an affirmative answer to this query, inasmuch as the British state's dependence upon the industry and its backers has been accentuated by the policy commitments discussed in the first two sections of this chapter. It is my contention, in other words, that the forces which conjoined to lend such a sense of urgency to offshore developments did more than simply generate a policy oriented to haste. This they certainly did, and, as we shall see in chapter 5, they thereby encouraged a pace of activity which both outstripped the legislative response with regard to safety by some distance and helped to generate an administrative structure geared to rapid development as much as to safety regulation; but they also spawned a particular kind of relationship within which the symbiotic balance between controllers and controlled became over-weighted in the latter's favour. While discussion of the way in which this adverse balance tipped the scales in relation to the safety regime will again be deferred to a later stage, it is important here to map out the underlying dimensions along which this dependent relationship became established.

The relationships which prevail between multinational companies and the host states within whose boundaries they operate have, of course, been characterized in different ways. Almost invariably, however, the terminology which is employed suggests some degree of co-operation or interdependency, coupled with conflict and antagonism. Speaking of such relationships in general, for example, Martinelli stresses the 'contradictions' between host countries and multinationals, the former deriving definite advantages as well as handicaps from the link. More specifically, he suggests, international capital 'needs' the host Government 'to perform a whole set of social and economic functions' such as defending property rights, creating necessary infrastructures and maintaining economic control and social peace in the country as a whole; at the same time, however, the fulfilment of these functions may be seriously impeded by the very operations of the multinationals themselves (for example, by their control over the terms of trade and financial flows).[136] While these adverse effects may be felt most acutely in Third World or developing countries, moreover, they are by no means restricted to such settings, as is indicated by Stuart Holland's incisive attack on 'academic agnosticism' with respect to the relative costs and benefits accruing to a country like Britain from multinational operations.[137] Similarly, the ambivalent nature of the relationship between the energy industry and Governments in the Western world has been admirably captured in Øystein Noreng's characterization of it as one of 'antagonistic interdependence':

> On the one hand, the Governments possess substantial energy resources, but in most cases they lack the relevant expertise and technology, and are reluctant to put public capital into high-risk ventures. On the other hand, the energy industry has the relevant expertise and technology, and in most cases is more willing than Governments to risk capital in new ventures, but it lacks the ownership of the resource potential. Consequently, Governments and the energy industry find themselves in a bargaining situation. Neither side can do without the other, but they have diverging interests, perceptions and demands.[138]

As the above description suggests, two crucial qualifications must be underlined before focusing attention on the way that the commitment to speed accentuated the host state's dependence on the international oil industry. In the first place, it is important to stress that however much the industry's bargaining power was enhanced by the position into which successive Governments got themselves, the relationship retained its symbiotic character, since the companies did still require Government to perform certain essential functions. Symbiosis does not necessarily involve exact equivalence in terms of reciprocal exchange. Nor, to take the second caveat, do they inevitably imply harmonious relationships or total identity of interest. Thus, despite what I will insist was the heightened dependency of the host state, divergent interests were perceived and pursued by Governments, often at the cost of considerable acrimony. In short, the relationship never deteriorated into a completely one-sided affair in which the authorities' role was that of abject surrender, a point which must be emphasized if we are to avoid a naive interpretation couched in terms that suggest the unresisted rape of the North Sea by multinational companies and international capital.

The first of the above qualifications can be fleshed out fairly briefly. From the industry's point of view, the most basic level at which there was a necessity for Government to play a facilitating role was in the creation of a legal framework for the conduct of operations on the Continental Shelf. Property rights in this instance not only had to be defended; they effectively had to be created in the first place. Until this was done, by virtue of the Continental Shelf Act of 1964,[139] private economic activity could not get under way in what was a legally novel situation.[140] Novel legal problems also arose in the context of the sometimes massive loans which companies, particularly but not exclusively the smaller licensees, required for oil-field development, since any loan which depended for its security upon a particular project itself was vulnerable to the possibility, however remote, that the relevant licence might be revoked. Thus, Government again provided a vital service in devising a form of undertaking which 'avoids fettering the Secretary of State's discretion, but provides banks with the necessary assurance of the continuity of their security'.[141] Similarly, banks were enabled to seek specific assurances about the future use of powers to control production under the licence.[142] North Sea investment pays off quickly and well once production commences – the Ninian field, for example, was at one point predicted to pay for itself in about three and a half years[143] – but it is very heavily front-loaded, and lenders and companies alike are therefore very sensitive to any delays or cut-backs in production. More generally, the role of Government in resolving legal difficulties *vis-à-vis* project finance was acknowledged by the Committee to Review the Functioning of Financial Institutions in 1978: 'in a new situation and in the absence of case law, ad hoc solutions to legal problems have had to be found; this has been done with the assistance of the UK Government which has adopted a positive attitude towards providing the necessary legal consents.'[144]

If the requirement of a legal framework and of arrangements to facilitate financing are two of the most obvious technical contexts in which it is not unreasonable to view the industry as symbiotically beholden to Government, a more general element of dependency can be discerned in the high value which its decision makers place upon predictability and political stability. As one commentator states, reviewing the slightly paradoxical situation in this connection, 'although oil men work in the world's biggest risk business, they hate political uncertainty.'[145] Thus it has been said that one of the UK Continental Shelf's greatest attractions was that companies could depend upon Governments of either hue to play more or less by the rules of the game,[146] an expectation which has arguably proved justified, protestations to the contrary and increased Government strictness in areas like taxation and participation notwithstanding.[147]

Nor should we forget the impact of an era in which the security of the companies' supplies has itself become increasingly tenuous, with the result that maximal dependence on areas where it is least likely to be jeopardized is at a premium. As Nore remarks of the Norwegian case, one side of the 'exchange' is that the companies 'gain access to secure supplies of high-quality oil close to the major markets in a geographic area outside the control of the OPEC cartel'.[148] In similar vein, the importance which the industry attaches to political stability has frequently been pointed out, countries like Britain and Norway once again being in the position of having something to offer which the companies increasingly need. According to one senior British official, speaking in the early days of oil development on the UK Shelf, it is 'absolutely right that when the pressure has been put on the international industry and the screws turned in other parts of the world, they are more interested in politically stable areas'.[149] From the other end of the chronological continuum, the same point has been echoed more recently by Mr Hamish Gray, Minister of State at the British Department of Energy:

> I think the sort of statement made by Mr Raisman [chairman of Shell UK] confirms that major oil companies are prepared to make investment in the North Sea, despite the unfriendly climate, for one reason more than any other, and that is political stability. And the sad scene which we have in the Middle East today, with Iran and Iraq, confirms more than ever that political stability, if you're making a multi-million pound investment, is all-important.[150]

Justice to the record and to those involved demands that the second qualification referred to earlier – the recognition and pursuit of divergent interests by Government – must also be spelled out in a little detail. To start at the beginning, for example, it is only fair to record that, whatever its other attractions, the discretionary system of licensing was also favoured because it permitted some degree of positive discrimination in favour of British interests. While efforts in this direction may not have been notably successful, particularly in the early years, they nonetheless attest the absence of any automatic assumptions about the advantages to be derived from unconditional largesse. Equally, the emphasis which was placed on the vigour of projected work programmes betokened not only an abiding concern for speed, but also a healthy suspicion that simply handing the enterprise over to the industry and to market forces might not necessarily even secure the achievement of politically defined objectives. Although Professor Dam and others have argued energetically that the logic of the discretionary system was not flawless, its utilization still stands witness to a relationship involving less than unqualified trust.[151]

Under the Heath administration of 1970–4, the divergence between public and private interest became both more pronounced and more overt. As we have already seen, for example, the revelations of the Public Accounts Committee in early 1973 amounted to a fairly damning indictment of both previous policies and of the extent to which unfettered pursuit of company interests would produce precious little by way of benefit to the British economy. Moreover, this report made it virtually inevitable that what could only be bitter negotiations about increased control and revenue 'take' would soon occur whichever party was in power. Indeed, it would be no exaggeration to say that at this point the gap between the 'national interest' and the industry's focal concerns was coming to be recognized as almost egregiously offensive on all political sides. While the impetus to do something about it may therefore have owed not a little to that fine political eye which is always looking to the broader ideological question of legitimacy, this does not detract from the fact that open acknowledgement was increasingly being accorded to the yawning chasm between the respective interests. Nor was the latter narrowed any when BP and Shell refused preferential treatment to the UK during the Arab oil embargo of late 1973. It would have taken a brave or even foolhardy politician to have repeated at this stage the bland assurances of those who, almost a decade earlier, had thought it 'only fair to say that the oil companies in the United Kingdom are very responsible bodies and that the first matter which they would wish to take into account would be the public interest'.[152]

When Labour returned to power in 1974, the task of securing more effective guarantees of what was perceived to comprise this public interest was given considerable priority. Again, the point here is not to suggest that the new administration was outstandingly successful on this score – indeed, my later argument will tend rather to support the converse – but to stress that the problem was recognized and efforts at remedial action made. In the face of considerable hostility and resentment, for example, changes were effected in both the taxation and participation regimes, while, as we have seen, a policy for depletion was laid down, however ambivalently. Some measure of the fraughtness of the relationship during these years can be gleaned from one description of Lord Kearton (chairman of BNOC) and Tony Benn, Eric Varley's successor at the Department of Energy, as two of the three men who 'have particularly incurred the wrath of the oil men'. According to the same author, the very name of the third, Peter Odell, is guaranteed to produce apoplexy on account of his repeated assertions that the companies wish only to cream off the most profitable oil, to the detriment of the national interest.[153] That the latter might not come very high in the industry's scheme of things was openly conceded in the context of the depletion debate during 1978:

> Commercial organizations tend to seek higher rates of return and more rapid pay-off periods than Governments, are often subject to pressures on capital and management resources, set a higher value than Government on the risks attaching to delay and have to budget for reinvestment elsewhere. For these reasons there may well be divergence between a depletion policy which Governments pursue in the national interest and that preferred by commercial operators.[154]

Although it is obviously too early to assess the record of the present Conservative Government in terms of any efforts to promote the public as opposed to the private interests associated with North Sea oil, there is no reason to believe that, doctrinaire political attitudes notwithstanding, there will be a complete *volte face*. As we have already seen, tentative noises have been made about depletion control, and certainly, Sir Geoffrey Howe has not hesitated to make opportunist use of windfall profits to increase taxation and government revenues. Similarly, despite plans to curb the role of BNOC, these have noticeably stopped short of full denationalization, doubtless in part because of the opportunities for 'discreet control' that it affords.[155] Nor should one fall into the tempting trap of assuming that because this is an administration committed to fostering the interests of private enterprise, relations with the offshore oil industry have therefore been consistently amicable. Initially, it is said, the industry was overjoyed at the prospect of a Tory Government pledged to restoring 'the confidence of the oil industry in the fair dealing of British Governments'; but after successive changes in the level and timing of tax payments, this attitude changed to one of anger, as the companies came to see themselves as being treated like a 'milch cow' to be milked 'every time they run short of a billion or two'.[156]

The qualifications, then, are entered: I am not asserting that the story of North Sea oil has simply been one of a gigantic 'rip-off' perpetrated by a totally autonomous industry against a succession of helpless, uncomplaining and incompetent British Governments. This said, however, it is also the case that, as in all symbiotic relationships, the host state has clearly become reciprocally dependent upon the industry and its backers in a number of significant ways. As suggested earlier, moreover, the extent of this dependency has been accentuated by the persistent preoccupation with speed which was discussed in earlier parts of this chapter. More specifically, the point here is that not only was speed adopted as a policy by successive Governments, but also that it increased the necessity for the co-operation of the international oil industry, banking and the capital markets (including London) in mustering the requisite operational capacity, finance and expertise to undertake the task of exploring and developing the North Sea within such a relatively restricted timespan.

On the operational front, it was precisely this need for haste that was consistently deployed to justify the extent of North Sea participation allowed to overseas interests and particularly to American companies. BP and the partly British Shell, even with the assistance of the Gas Council and the National Coal Board, would clearly have been incapable of mounting such a massive operation on their own, or at least not at the pace deemed desirable. In consequence, substantial outside involvement was seen as inevitable and, however ambivalently, to be welcomed. As Sir Robert Marshall succinctly explained to the Public Accounts Committee late in 1972, 'If you go very fast and there is a highly developed international industry and you are only a small proportion of that industry, then you have to make big calls on other people.'[157] Thus, after correcting its arithmetic to remedy errors which had both attributed almost twice its actual stake to the British public sector and counted Shell as 100 per cent British,[158] the Department of Trade and Industry calculated that by the beginning of 1973, the British share in licensed territory was 32 per cent.[159] Similarly, and by using a method of calculation which gave results twice as optimistic as those produced by the International Management and Engineering Group,[160] it put the British contribution to supplies and services at 50 per cent, a figure it claimed was 'not bad',[161] Once again, the justification was speed:

> The bulk of the oil industry is foreign. Next to the Americans this country has the largest single stake in it, but simply following the great emphasis of American enterprise in the petroleum industry established across the world, American suppliers are the most prominent and it was quite clear that the faster one went and the less notice

that was given towards exploration and exploitation plans in the North Sea, the less the immediate share of British industry.[162]

From the outset, then, the penetration of overseas interests into the North Sea oil industry was substantial, and as we shall see, this was to have a constraining effect on Government when it did belatedly get around to tightening up controls from 1974 onwards.[163] Moreover, while the picture has shown some change since then – particularly with the British share of goods and services rising to 79 per cent by the end of 1979[164] – this should not be exaggerated. Although the establishment of the BNOC was floated on the ideology of majority public participation, this initially turned out to be, in the much-quoted words of the *Banker,* 'merely an option to buy oil . . . and an assurance that most of the refining is done in the UK'.[165] As far as exploration and production were concerned, the new corporation had a long way to go in building up its own operational capacity, and by the end of 1980 it still only controlled around 8 per cent of North Sea production.[166] In contrast, American companies controlled 54 per cent of production at the beginning of 1978.[167]

On the financial front too, the speed of development increased the degree of dependence upon the international industry and its financiers. As the Committee to Review the Functioning of Financial Institutions again pointed out, 'Exploitation of North Sea oil reserves depended upon the successful completion of a massive programme of capital investment, much of it compressed within a short space of time.'[168] Nor was this any exaggeration. By the beginning of 1972, it was estimated that £450–£500 million had thus far been spent on exploration and production,[169] but these figures were soon to be dwarfed as activity came to centre increasingly on the more northerly areas and as exploration led on to the much more costly business of development. Between 1976 and 1979, for example, capital investment was running at over £2 billion per annum, and total offshore-related investment was reckoned to be equivalent to 20–25 per cent of annual UK industrial investment;[170] by the end of the latter year, cumulative investment in the Continental Shelf enterprise was being put at £18 billion in 1979 prices.[171]

One consequence of such vast capital requirements over a comparatively short period was that the exercise came to hinge crucially upon the capacity of major oil companies to finance ventures from their own internally generated funds. Shell and Esso, for example, together account for around 35 per cent of North Sea expenditure and have largely managed to meet their capital requirements in this way.[172] But the 'daunting' scale of the sums required sometimes even outran the resources of the majors, and they were forced to join their lesser brethern in seeking external finance on a fairly substantial scale.[173] In 1970, the Chase Manhattan Bank noted a pronounced trend in this direction within the international oil industry as a whole, and indeed some two years later fears were even being expressed that the companies might be reaching the limits of their outside borrowing capability. Such a scenario, it was suggested, augured ill for the development of more costly offshore discoveries unless profit margins could be widened.[174]

Despite price increases and 'windfall' profits over the remainder of the decade, external borrowing continued as operations moved into more difficult areas and as projects entered the more costly development phase. In 1972, for example, BP borrowed $468 million and £180 million for the development of its Forties field, while Occidental raised $325 million in syndicated loans for its Piper and Claymore interests between 1974 and 1976. Shell had to find £250 million for the development of Dunlin, Brent and Cormorant, while BNOC's obvious dearth of internal funds from other operations led it to borrow $825 million to finance its growing North Sea activities in 1977. By the end of that year, *identifiable* project loans alone amounted to $3100 million and £550 million, figures which corresponded to around 25 per cent of the total estimated capital expenditure for the eighteen fields then existing.[175]

In the event, the financial system 'proved equal to the challenge' posed by such demands.[176] But once again, this was only accomplished by dint of substantial outside assistance. Writing in 1975, MacKay and Mackay noted that the sums of money involved were far in excess of what the domestic capital market or, indeed, Eurodollar and European Community funds could provide, with the inevitable result that US finance was already dominant and likely to remain so.[177] Although much of the requisite lending was probably, in the end, organized through the British financial system,[178] it was by no means predominantly from British sources. According to the Committee to Review the Functioning of Financial Institutions, over 60 per cent of all the identifiable loan finance outstanding or committed by mid-1977 had been raised from overseas, and particularly American, banks.[179] Thus, in addition to dependence upon the capital invested by the major oil companies from their own funds, the development of Britain's Continental Shelf has leaned heavily on what the above Committee, with a nicely anthropomorphic touch, called 'the contribution of non-residents to North Sea oil financing'.[180] The point was put dramatically by one oil correspondent in 1978:

> If the UK had had to go it alone – in the way that British companies built the railway systems of the mid-nineteenth century in this country, which is after all a comparable venture – then the strain on the financial system might have been much greater.
>
> But the key to the financing of the North Sea has been the involvement from the beginning of the established international oil companies, who have been large enough to treat North Sea exploration and development as part of their world-wide activities. . . . Further, so dominant has been London as a centre of the international capital market, that UK banks, or international banks with branches operating aggressively through the international capital markets in London, have over the years developed a variety of techniques that have enabled them to put together financial packages to cope with the special demands of large-scale oil development. . . .[181]

But it was not only in terms of operational capacity and the industry's capacity to raise the requisite finance that the rapid development of Britain's offshore resources involved a marked degree of dependency, for the authorities were also heavily reliant upon the industry in another crucial respect. Although one of the world's largest oil companies is nearly half British and another is wholly so, there is no doubt that when the mineral resources of the North Sea first came on to the political agenda, UK officials found themselves at a severe disadvantage with regard to information and expertise. Technological hegemony, of course, is one of the key factors in multinational survival, and the North Sea oil industry was to prove no exception, while on the less sophisticated plane which principally concerns us here, the potential of the UK's Continental Shelf was a largely unknown quantity. Whatever the industry may have known about its possibilities, and there have been some suggestions that this was more than it was telling at the time,[182] officialdom was working substantially in the dark.

Nor were the authorities particularly well placed to remedy this situation, at least to begin with. As the Department of Trade and Industry later recalled in discussing the predicament of its predecessor in this respect, the Ministry of Power could not even demand information on the costs or results of the operations which were taking place 'as of right on the high seas' at the beginning of the 1960s. Indeed, the need to overcome this disadvantage was one of the primary

bureaucratic motives for preparing the legislation which became the
Continental Shelf Act of 1964.[183] Even when the gap was plugged in
relation to 'physical information' on results, moreover, serious doubts
still remained as to the Ministry's capacity to evaluate what it was
being told. A chronic lack of technical staff persisted up to and beyond
the end of the decade, and seismic and drilling results were passed to
the Institute of Geological Sciences for assessment. Subsequently
described in less than flattering terms as 'an academic institution with
limited expertise in petroleum production',[184] this body's funding was
grossly inadequate for the task in hand up to 1967 at the very
earliest.[185]

As already suggested, part of the problem here was, of course, that
the North Sea enterprise was new. But the difficulty was greatly
exacerbated by the almost 'Catch 22' situation created by an official
policy which, for reasons outlined at more than one previous point in
this chapter, favoured speed. On the one hand, successive Governments wanted to find out as rapidly as possible what kind and scale of
asset was now available to them, and for this purpose wished to
encourage speedy exploration by the industry which alone could furnish
the answer; on the other hand, the more successful they were in this
endeavour, the less the time available for gearing the administration up
to the task of either handling the information supplied or developing
its own independent monitoring capacity. Once again, the evidence
given to the 1972 Public Accounts Committee by the Permanent
Secretary at the Department of Trade and Industry was riven with
precisely this dilemma. Time and again, he reiterated his belief that the
speed of exploration to date had been perfectly justified because of the
need to know as quickly as possible about the nature of this valuable
national asset.[186] At the same time, he had to concede, among other
things, that the problems confronting the Institute of Geological
Sciences stemmed from the fact that 'this was an operation which went
off with tremendous pace and . . . that particular Office was not set up
for it then.'[187] In more general terms, his account of the authorities'
capacity to handle the early stages of North Sea exploration provoked
some blunt and almost rhetorical questioning from his examiners:

> Would it therefore be reasonably true to say that in contemplating
> the exploitation of this massive new piece of publicly owned area we
> did not take time to gear ourselves in terms of the information that
> our Government Department ought properly to acquire about the
> potentialities of the area in terms of funding and in terms of know-
> how available directly to the government? We were not geared to do
> it?[188]

By the time that this question was posed, late in 1972, civil servants were able to claim that substantial progress had already been made on covering the information deficit. For all the inadequacies of the old Ministry of Power and the Institute, for example, it had been felt that departmental and ancillary knowledge was sufficient to mount a major policy review in 1968–9,[189] and by the following year the number of technical staff had doubled, albeit only to six. This expansion continued under the auspices of the Department of Trade and Industry, the number of technical experts rising to nine by 1972.[190] Moreover, this Department could even claim that in some respects it occupied a position of considerable informational advantage, 'remembering that all seismic information and so forth obtained under licence must be revealed to the Government so that the Government is the major repository of this information and has more information than any single company or group operating'.[191] With the subsequent establishment of the Department of Energy and the BNOC, potential access to requisite information was enhanced still further. 'Part of the whole purpose' in setting up the latter organization, for example, was 'to improve the information available to the Government about the oil industry' and, more specifically, to implement the 'justifiable' claim that there should be 'more access to information about the activities of the licensees'.[192] As for the Department of Energy, its capacity to acquire information of a quite detailed kind became increasingly substantial:

> As more and more oil finds were made through the 1970s, so the Department refined its procedures. Once a discovery has been made and a company wishes to develop, a wide range of information has to be supplied to the Department. There are two schedules (A and B) of requirements. Stringent programmes giving details of field installations, wells, other relevant works, the quantities of petroleum to be produced and the construction schedule must be submitted. For their Development and Production Programme, the Department of Energy requires background information to include an historical introduction to the field, details of the area to which the programme will relate, geological and geophysical data, formation parameters, reservoir fluid parameters and reserves, development and production plans, profiles, transportation of petroleum, pollution-prevention measures, shore terminals, costs, construction schedule, process flow diagram and other matters.[193]

Although these developments all reflected a growing administrative capacity to collect and assess data about North Sea operations, the continuing difficulties in this respect should not be underestimated. As late as 1973, for example, information on costs as opposed to results was still being supplied on a voluntary basis, and even if some officials felt that this system operated fairly satisfactorily, they could not deny that it did nonetheless place them at a certain disadvantage.[194] Nor would they perhaps have been quite as satisfied if they had been privy to the complaints of some of their colleagues who, around the same time, were apparently citing instances in which requests for information not statutorily required were being turned down.[195] Equally, even after the emergence of the Department of Energy, doubts still remained with regard to the effectiveness of its 'handling of drilling and other information given to it by the companies and the quality of input it, in turn, feeds into policy decisions'.[196] While BNOC is said to have generated intense dislike among the companies precisely because it was 'dispelling the mystique about their knowledge',[197] its own communication of general information and plans to the Energy Department has been described as 'sporadic'.[198] In more general terms, as recently as the end of 1980, the Minister of State at the Department was still lamenting its short supply of 'experts who are trained in the relevant geological and geophysical sciences and in oil and gas engineering', personnel who were needed 'to advise on oilfield developments and operations more generally'.[199] As we shall see in the next chapter, moreover, the way in which the shortfall in expertise had been made up over the years was only to raise further questions about officialdom's dependence on the industry in this context. Above all, increased expertise and information flow did not resolve one perennial problem, namely, that Governments intent upon establishing the full scale of their offshore reserves as quickly as possible were still substantially dependent upon the international oil industry to supply the answer.

The most obvious effect of the heightened dependency which I have described in this section was that even when British Governments did start to assert themselves, particularly after 1974, they were constantly confronted with threats from an industry whose bargaining power was greatly augmented by their own concern for speed. Like its nineteenth-century forebears, the international and multinational capital of today always has the ultimate sanction of invoking the old maxim that 'like love, its workings must be free as air; for at sight of human ties, it will spread the light wings of capital and fly away from bondage.'[200] Less graphically, the history of the North Sea in the 1970s is littered with industry warnings and reports, often felicitously timed to coincide with new government measures, and usually amounting in effect to threats of a decline in drilling activity or of a fall-off in investment. Steps in this latter direction do, for example, seem to have greeted participation proposals in 1974,[201] while four years later, a Labour request for an investigation into declining drilling work was pre-empted by claims that the slow-down was attributable to its own policy on tax and other matters.[202] Early in 1979, planned changes in petroleum revenue tax were said to be curtailing future development plans and hitting exploration, with the additional twist that only large or low-cost fields might be developed in the future. Such an outcome, added the United Kingdom Offshore Operators Association with characteristic bluntness, would lead to an overall loss in government revenues.[203] In the following year, a new Government which had hoped to hasten the search for oil was being told by the Association's director-general that delays in announcing new licences, the imposi-tion of constant changes in petroleum revenue tax and the declared intention of imposing a supplementary tax were all contributing to a continued down-turn in activity.[204] When the last of these taxes was confirmed in the 1981 Budget, he explained that it was not so much a threat as a fact of life to point out that increased taxation on this scale might make other oil-producing areas of the world more attractive than the North Sea.[205]

While the facts of life, threats or warnings may not have deterred Governments from stepping up their tax 'take' – the price/profit spiral in oil made such steps easier to justify and harder to resist[206] – there is little doubt that the heightened dependency of the host state has had a constraining effect upon efforts to impose a tougher operating regime on the offshore oil industry. Thus, to take the most obvious example, however much the extent of overseas penetration bolstered the case for Labour's planned increase in state participation, it also set limits upon the extent to which such a policy could be carried through. With 60 per cent of Britain's offshore reserves in the hands of foreign companies, outright nationalization would have lent embarrassing international ramifications to the issue of compensation, a matter not at all simp-lified by the possibility of US retaliation against BP's considerable stake in the Alaskan oilfields.[207] Moreover, as Adrian Hamilton points out, the broader background against which Government undertook what therefore had to be voluntary negotiations was one in which 'the Treasury placed increasing reliance on the North Sea potential oil flows and on American favour in order to raise its large international loans to finance its budget deficit and the continuing deficits on the balance of payments.'[208] Hence, while control over licensing and assignment was indeed used to pressure some of the less substantial companies into 'voluntary' agreements on participation, many of the majors were able to hold out for a better deal than originally envisaged. In consequence, the terms on which BNOC would become involved in existing fields were progressively modified to the point where state participation became a more than slightly ambiguous entity which has been variously described as 'a purely paper financial transaction', an arrangement of largely 'symbolic' significance and a 'purchase agreement'.[209]

While the details of these and other modifications need not concern us here, it is important to stress that they comprised part of a wider pattern of attenuation that was apparent right from the beginning of the period which saw the first serious efforts at more effective control. At the very outset, Labour had described the coming task as being 'to work out a new structure and taxation arrangements which would combine their manifesto commitment with an assurance to the oil companies about the substantial role that would remain to them'.[210] Similarly, the assumption of powers to control depletion was quickly coupled not only to a reassertion of rapid development as the immediate objective, but also to an assurance 'that our depletion policy and its implementation will not undermine the basis on which they [the companies and banks] have made plans and entered into commitments'.[211] When Lord Balogh, one of the industry's most vociferous critics, came to announce concessions in the Petroleum and Submarine Pipelines Bill in November 1975, he likewise adopted a highly placatory and significant tone. 'The Government', he insisted, 'values the presence of oil operators far too much to damage their confidence.'[212] The strenuous efforts made to avoid such a contingency are evident in the extent, considerable by any standard, to which the industry was brought into negotiations on a wide range of issues. As Cameron astutely remarks, 'It is clear that the terms of existing licences were revised as the result of a unilateral decision taken by the British Government, but that they were not revised *unilaterally*.'[213]

Conceived at a time when the issue was at its most politicized, this deferential approach to controlling the offshore oil industry has survived into the present and has continued to index the dilemma confronting Governments in this context. While they may genuinely seek to secure better guarantees of the public interest, they cannot pursue this objective with a degree of resolution that might risk alienation of the companies. That some measure of success should have been achieved in the former context is not to be denied; that more should not have been accomplished is largely attributable, I suggest, to the unbalanced nature of the relationship which, against the background outlined in the first two sections of this chapter, has grown up between the host state and the industry over the past fifteen years. While certainly symbiotic, this relationship has involved Governments which are caught between their own commitment to speed, hostages already surrendered to fortune and their concomitantly heightened dependence on the oil industry. To the effects of this dependence, as of speed itself, upon the regulation of safety we will return in the next chapter.

Notes and References

1. D. Purdy, 'British Capitalism since the War', *Marxism Today*, vol. 20, 1976, p. 271.
2. B. Jessop, 'The State in Post-War Britain', in R. Scase (ed.), *The State in Western Europe*, London, Croom Helm, 1980, p. 37.
3. S. Strange, *Sterling and British Policy*, London, Oxford University Press, 1971, p. 323.
4. Purdy, 'British Capitalism since the War', p. 316.
5. F. Longstreth, 'The City, Industry and the State', in C. Crouch (ed.), *State and Economy in Contemporary Capitalism*, London, Croom Helm, 1979.
6. ibid., p. 177.
7. ibid., p. 184.
8. Purdy, 'British Capitalism since the War', p. 316.
9. Jessop, 'The State in Post-War Britain', p. 40. According to Susan Strange, one of the characteristics of the period was 'the loss of power by other Government Departments to the Treasury, and by other Ministers to the Chancellor of the Exchequer; the influence of the Bank of England on the Treasury; and the loss of power by Parliament to both'. *Sterling and British Policy*, p. 301.
10. Strange, *Sterling and British Policy*, ch. 8.
11. A good review of these developments is to be found in P. Odell, *Oil and World Power*, Harmondsworth, Penguin Books, 1979. Much of the following discussion is based on this source.
12. D. I. MacKay and G. A. Mackay, *The Political Economy of North Sea Oil*, London, Martin Robertson, 1975, p. 5.
13. C. Robinson and J. Morgan, *North Sea Oil in the Future*, London, Macmillan, 1978, p. 191.
14. See, for example, J. Stork, *Middle East Oil and the Energy Crisis*, New York, Monthly Review Press, 1975, p. 60.
15. Odell, *Oil and World Power*, p. 112.
16. Libya, for example, was determined to avoid such monopolisation by the majors during the 50s. A. Sampson, *The Seven Sisters*, Sevenoaks, Hodder & Stoughton, 1976, p. 159.

17. Odell, *Oil and World Power*, p. 15.
18. Ø. Noreng, *The Oil Industry and Government Strategy in the North Sea*, London, Croom Helm, 1980, p. 37.
19. Odell, *Oil and World Power*, p. 113.
20. See below, p. 96.
21. Noreng cites 'growing pessimism about expanding reserves in traditional cheaper areas of production' in this respect *The Oil Industry*, p. 75. Other possible explanations are discussed in P. Cameron, 'Property Rights and the Role of Government: the Problem of Property in North Sea Oil', Edinburgh, unpublished Ph.D. thesis, 1980, pp. 91–4.
22. The Committee of Public Accounts, 'North Sea Oil and Gas', *Parliamentary Papers*, 1972–3, vol. 14, London, HMSO, 1973, p. 24.
23. Odell, *Oil and World Power*, p. 114.
24. Committee of Public Accounts, 'North Sea Oil and Gas', p. 24.
25. P. Nore, *Six Myths of British Oil Policies*, London, Thames Papers in Political Economy, 1976, p. 14.
26. A. Hamilton, *North Sea Impact*, London, International Institute for Economic Research, 1978, p. 17.
27. Committee of Public Accounts, 'North Sea Oil and Gas', p. 25.
28. ibid., p. 24.
29. *Parliamentary Debates* (Commons), vol. 688, 1963–4, 28 January 1964, col. 218.
30. *Parliamentary Debates* (Commons), vol. 692, 1963–4, 7 April 1964. col. 897.
31. Committee of Public Accounts, 'North Sea Oil and Gas', p. 45.
32. Ministry of Power, *Fuel Policy*, Cmnd. 3438, London, HMSO, 1967, p. 43.
33. Robinson and Morgan, *North Sea Oil in the Future*, pp. 21–2.
34. Hamilton, *North Sea Impact*, p. 24.
35. Ministry of Power, *Fuel Policy*, p. 42.
36. S. Brittan, *Steering the Economy*, Harmondsworth, Penguin Books, 1971, p. 379.
37. ibid., pp. 376–7.
38. ibid., p. 396.
39. Jessop, 'The State in Post-War Britain', p. 40.
40. Committee of Public Accounts, 'North Sea Oil and Gas', p. 29.
41. E. Mandel, *The Second Slump*, London, Verso Books, 1980, p. 10; I. Gough, 'State Expenditure in Advanced Capitalism', *New Left Review*, vol. 92, July–Aug. 1975, p. 88.
42. R. Skidelsky, 'The Decline of Keynesian Politics', in Crouch (ed.), *State and Economy in Contemporary Capitalism*, p. 74.
43. Brittan, *Steering the Economy*, p. 383. For a useful discussion of the collapse of the Bretton Woods system, see A. Shonfield, *International Economic Relations of the Western World 1959–1971*, London, Oxford University Press, 1976, ch. 3.

44. Odell, *Oil and World Power*, p. 216; see also J. Chevalier, *The New Oil Stakes*, London, Allen & Unwin, 1976; and T. Rafai, *The Pricing of Crude Oil*, New York, Praeger, 1974.
45. Sampson, *The Seven Sisters*; Stork, *Middle East Oil and the Energy Crisis*.
46. Ministry of Power, *Fuel Policy*, p. 24.
47. P. Nore, 'Oil and Contemporary Capitalism', in P. Nore and F. Green (eds.), *Issues in Political Economy*, London, Macmillan, 1979, p. 115.
48. A. Glyn and J. Harrison, *The British Economic Disaster*, London, Pluto Press, 1980, p. 90.
49. P. Donaldson, *Economics of the Real World*, Harmondsworth, Penguin Books, 1978, p. 104.
50. A. Prest and D. Coppock (eds.), *The U.K. Economy*, London, Weidenfeld & Nicolson, 1980, p. 311.
51. Glyn and Harrison, *The British Economic Disaster*, pp. 75–6.
52. Prest and Coppock (eds.), *The U.K. Economy*, p. 306.
53. The importance of the relative strength of the British working class in this respect is stressed by many writers on Britain's post-war decline. Usually, its strength at the workplace is emphasised. See, for example, Purdy, 'British Capitalism since the War'.
54. P. Nore, 'International Oil in Norway' in J. Faundez and S. Piccioto, *The Nationalisation of Multinationals in Peripheral Economies*, London, Macmillan, 1978, p. 190.
55. Ministry of Power, *Fuel Policy*, p. 36; MacKay and Mackay, *The Political Economy of North Sea Oil*, p. 4.
56. Committee of Public Accounts, 'North Sea Oil and Gas', p. 32.
57. ibid., p. 64.
58. ibid., p. 31.
59. ibid.
60. ibid., p. xxiii.
61. D. Howell, 'North Sea Oil: a Reply to Lord Balogh', *Banker*, March 1974, p. 563.
62. For figures on North Sea production and inland energy consumption from 1974 onwards, see Department of Energy, *Development of the Oil and Gas Resources of the United Kingdom* (henceforward *The Brown Book*), London, HMSO, 1975–80.
63. Department of Trade and Industry, *Production and Reserves of Oil and Gas on the United Kingdom Continental Shelf*, London, HMSO, 1973, p. 1.
64. Stork, *Middle East Oil and the Energy Crisis*, p. 224.
65. *Parliamentary Debates* (Commons), vol. 867, 1973–4, 9 January 1974, col. 135.
66. C. Allsopp, 'The Management of the World Economy' in W. Beckerman (ed.) *Slow Growth in Britain*, Oxford, Oxford University Press, 1979, p. 147.

134   *The Political Economy of Speed*

67. See, for example, P. Ormerod, 'The Economic Record', in N. Bosanquet and P. Townsend (eds.) *Labour and Equality*, London, Heinemann, 1980, p. 49.
68. Mandel, *The Second Slump*, pp. 17ff.
69. Glyn and Harrison, *The British Economic Disaster,* p. 100.
70. Mandel, *The Second Slump*, p. 5.
71. Institute of Development Studies, *North Sea Oil: the Application of Development Theories*, Brighton, University of Sussex, 1977, p. 25.
72. C. Milner, 'Payments and Exchange Rate Problems' in P. Maunder (ed.) *The British Economy in the 1970s*, London, Heinemann, 1980, p. 235.
73. For a discussion of this vulnerability and its causes, see J. Metcalfe, 'Foreign Trade and the Balance of Payments', in Prest and Coppock (eds.), *The U.K. Economy*, pp. 153ff.
74. Purdy, 'British Capitalism since the War', p. 274.
75. ibid., p. 316.
76. Ormerod, 'The Economic Record', p. 61.
77. Probably the best available review of this period is to be found in Bosanquet and Townsend (eds.), *Labour and Equality*, while relevant statistics are systematically set out in Prest and Coppock (eds.), *The U.K. Economy*, and in *Annual Abstract of Statistics*, London, HMSO.
78. G. Arnold, *Britain's Oil*, London, Hamish Hamilton, 1978, p. 11.
79. ibid.
80. *Parliamentary Debates* (Commons), vol. 867, 1973–4, 9 January 1974, col. 135.
81. *Parliamentary Debates* (Commons), vol. 870, 1974, 20 March 1974, col. 1218.
82. Department of Energy, *United Kingdom Offshore Oil and Gas Policy*, Cmnd. 5696, London, HMSO, 1974, p. 5.
83. *Brown Book,* 1978, p. 53.
84. *Brown Book,* 1975, p. 36.
85. Department of Energy, *Production and Reserves of Oil and Gas in the United Kingdom*, London, HMSO, 1974, p. 2.
86. *Brown Book,* 1975, p. 36.
87. *Brown Book,* 1978, p. 53.
88. For details of these rounds, see *Brown Book,* 1979, p. 55.
89. Department of Energy, *Energy Policy – a Consultative Document*, Cmnd. 7101, London, HMSO, 1978, p. 36.
90. See, for example, Robinson and Morgan *North Sea Oil in the Future*, p. 22.
91. According to at least one report, however, he found this task 'fraught with problems'. *Sunday Times,* 6 July 1980.
92. For a discussion of possible reasons, see Cameron, 'Property Rights and the Role of Government'.
93. *Brown Book,* 1979, p. 20. Oil's share of the market had fallen from around 49 per cent in 1974 to some 44 per cent in 1978.

94. ibid., pp. 32 and 39. In addition, the Claymore field had run the complete gamut from discovery to production during Labour's period in office.
95. The only really major field involved was Statfjord, with estimated recoverable reserves of 412 million tonnes, but the British share in this was only 16 per cent.
96. The following figures are taken from the Department's annual reports for the period in question.
97. *Brown Book*, 1977, p. 8.
98. Metcalfe, 'Foreign Trade and the Balance of Payments', p. 162.
99. Cmnd. 7143, London, HMSO, p. 5.
100. *Sunday Times*, 15 January 1981.
101. Central Statistical Office, *Economic Trends*, No. 327, London, HMSO, 1981, p. 36.
102. ibid., p. 18.
103. *Guardian*, 9 February 1981.
104. *Sunday Times*, 20 July 1980.
105. *Financial Times*, 29 May 1980.
106. P. Forsyth and J. Kay, *The Economic Implications of North Sea Oil Revenues*, London, Institute of Fiscal Studies, 1980.
107. *Sunday Times*, 6 December 1980.
108. See, for example, *Observer*, 8 February 1981, where Adrian Hamilton and William Keegan put forward such a suggestion as part of a package for 'the road to recovery'.
109. *Scotsman*, 22 January 1981.
110. *Guardian*, 27 July 1979.
111. *The Times*, 11 November 1980.
112. *Financial Times*, 2 May 1980.
113. *Oilman*, 1 December 1979. Part of the problem in the latter respect was said to stem from the fact that BNOC would no longer automatically have a 51 per cent stake in all future licences.
114. *Financial Times*, 25 October 1980.
115. *Oilman*, 20 December 1980.
116. *Sunday Times*, 6 July 1980.
117. *Financial Times*, 3 October 1980.
118. *Observer*, 27 April 1980.
119. *Financial Times*, 15 March 1980.
120. *Financial Times*, 1 October 1980.
121. *Guardian*, 28 February 1981.
122. When BNOC's Clyde field (shared with Shell and Esso) became a candidate for a possible delay on development, the nature of the contradictory pulls in this context became quite apparent. On the one hand, the Treasury pressed for a five-year delay to ease pressure on the PSBR through capitalizing on the Corporation's surplus. On the other, the Department of Energy seems to have vacillated between abandoning the delay altogether, on account of reduced output projections

from the North Sea, and settling for a compromise postponement of two or three years. In the end, a two-year delay was agreed. *Sunday Times,* 26 October 1980; *Financial Times,* 23 April 1980; *Oilman,* 20 December 1980.

123. *Financial Times,* 15 March 1980.
124. ibid., 3 October 1980.
125. *Sunday Times,* 6 December 1980.
126. See, for example, the *Scotsman's* powerful editorial in response to the comments made by Mr John Raisman, the chairman of Shell UK, referred to above. *Scotsman,* 22 January 1981.
127. Reported in *Financial Times,* 1 October 1980.
128. Mr Merlyn Rees asked for Government assurances on this score in the course of a debate on the high cost of energy to British industry, in January 1981. *Scotsman,* 22 January 1981.
129. *Guardian,* 31 December 1980.
130. Robinson and Morgan, *North Sea Oil in the Future,* p. 44.
131. For a brief discussion of the literature surrounding this issue, see J. Scott, *Corporations, Classes and Capitalism,* London, Hutchinson, 1979, ch. 6.
132. Longstreth, 'The City, Industry and the State', p. 189.
133. Jessop, 'The State in Post-War Britain', p. 85.
134. Scott, *Corporations, Classes and Capitalism,* p. 94 and passim.
135. Noreng, *The Oil Industry and Government Strategy in the North Sea,* p. 211.
136. A. Martinelli, 'Multinational Corporations, National Economic Policies and Labour Unions', in L. Lindberg *et al.* (eds.), *Stress and Contradiction in Modern Capitalism,* Lexington, Lexington Books, 1975, p. 428.
137. S. Holland, *The Socialist Challenge,* London, Quartet Books, 1975, ch. 3.
138. Noreng, *The Oil Industry and Government Strategy in the North Sea,* p. 19.
139. See chapter 5 below.
140. For a good discussion of what was involved here, see Cameron, 'Property Rights and the Role of Government, chs. 1 and 2.
141. *Brown Book,* 1978, p. 27. In the case of BP's project loan for the development of the Forties field, an assurance was given that in the event of a default that would normally lead to revocation, the banks could take over the licence on exactly the same terms and conditions as other licensees. The Committee of Public Accounts, 'North Sea Oil and Gas', p. 91.
142. *Brown Book,* 1978, p. 27.
143. Committee to Review the Functioning of Financial Institutions, *The Financing of North Sea Oil,* London, HMSO, 1978, p. 6.
144. ibid., p. 10.
145. *Financial Times,* 15 March 1980.

146. Institute of Development Studies, *North Sea Oil: the Application of Development Theories,* p. 24.
147. See Cameron, 'Property Rights and the Role of Government', passim.
148. Nore, 'International Oil in Norway', p. 196.
149. Committee of Public Accounts, 'North Sea Oil and Gas', p. 79.
150. BBC News, 25 September 1980.
151. K. Dam, *Oil Resources,* Chicago, University of Chicago Press, 1976, ch. 4.
152. *Parliamentary Debates* (Commons), vol. 688, 1963–4, 28 January 1964, col. 256.
153. Arnold, *Britain's Oil,* p. 2.
154. *Brown Book,* 1978, p. 54.
155. *Observer,* 27 April 1980; *Financial Times,* 3 October 1980.
156. *Sunday Times,* 30 November 1980.
157. Committee of Public Accounts, 'North Sea Oil and Gas', p. 78.
158. ibid., p. 68.
159. ibid., p. 89.
160. ibid., pp. xxx and 121.
161. ibid., p. 118.
162. ibid., p. 121.
163. See below, pp. 120 ff.
164. *Brown Book,* 1980, p. 22.
165. See, for example, Arnold, *Britain's Oil,* p. 54.
166. *Financial Times,* 3 October 1980.
167. Arnold, *Britain's Oil,* p. 344.
168. Committee to Review the Functioning of Financial Institutions, *The Financing of North Sea Oil,* p. 1.
169. Committee of Public Accounts, 'North Sea Oil and Gas', p. 52.
170. See, for example, *Brown Book,* 1978, p. 27.
171. *Brown Book,* 1980, p. 19.
172. Committee to Review the Functioning of Financial Institutions, *The Financing of North Sea Oil,* p. 15.
173. ibid., p. 5.
174. Committee of Public Accounts, 'North Sea Oil and Gas', p. 57.
175. Committee to Review the Functioning of Financial Institutions, *The Financing of North Sea Oil,* p. 10. Details of project and other identifiable loans are given in appendix 2.2 and appendix 2.3 to the report.
176. ibid., p. 5.
177. MacKay and Mackay, *The Political Economy of North Sea Oil,* p. 76.
178. Committee to Review the Functioning of Financial Institutions, *The Financing of North Sea Oil,* p. 15.
179. ibid., p. 26.
180. ibid., p. 16.
181. *Oilman,* 18 November 1978.
182. See, for example, Nore, 'International Oil in Norway', p. 168.
183. Committee of Public Accounts, 'North Sea Oil and Gas', p. 123.

184. Hamilton, *North Sea Impact,* p. 31.
185. Committee of Public Accounts, 'North Sea Oil and Gas', p. 77.
186. ibid., p. 124.
187. ibid., p. 77.
188. ibid., pp. 77–8.
189. ibid., p. 27.
190. ibid., p. 35.
191. ibid., p. 67.
192. *Parliamentary Debates* (Commons), vol. 899, 1974–5, 5 November 1975, col. 407.
193. Arnold, *Britain's Oil,* p. 153.
194. Committee of Public Accounts, 'North Sea Oil and Gas', p. 126.
195. I. White *et al., North Sea Oil and Gas,* Norman, University of Oklahoma Press, 1974, p. 40.
196. *Financial Times,* 6 February 1975.
197. Arnold, *Britain's Oil,* p. 165.
198. Noreng, *The Oil Industry and Government Strategy in the North Sea,* p. 148.
199. *Parliamentary Debates* (Commons), vol. 991, 1980–1, 6 November 1980, col. 1478.
200. A. Ure, *The Philosophy of Manufactures,* London, 1835, p. 453.
201. Cameron, 'Property Rights and the Role of Government', ch. 3.
202. *Oilman,* 21 October 1978.
203. ibid., 24 February 1979.
204. *Sunday Times,* 30 November 1980.
205. BBC (Scotland), 11 March 1981.
206. Indeed, the imposition of a 20 per cent supplementary tax has been described as a gamble based on the prediction of further price rises, which will mean that North Sea investment may not be too hard hit.
207. For a discussion of this side to the matter, see Cameron, 'Property Rights and the Role of Government', ch. 3.
208. Hamilton, *North Sea Impact,* p. 42.
209. Hamilton, *North Sea Impact,* p. 117; Dam, *Oil Resources,* p. 121; Noreng, *The Oil Industry and Government Strategy in the North Sea,* p. 52.
210. Department of Energy, *United Kingdom Offshore Oil and Gas Policy,* Cmnd. 5696, London, HMSO, 1974, p. 3.
211. *Brown Book,* 1975, p. 36.
212. *Financial Times,* 12 November 1975.
213. Cameron, 'Property Rights and the Role of Government', p. 184.

5

The Growth of a Special
Regulatory Framework

The only thing that has clearly emerged from the debate in this
house and in another place is that not only do we not know what
we are doing, but nobody knows what is to be done.
Mr Leslie Hale, MP,
Continental Shelf Bill,
Second Reading Debate,
January 1964

Safety Second

Paradoxically, perhaps, it was the oil industry's symbiotic need for Government rather than the latter's need for the industry that first led to the question of offshore safety becoming a live legislative issue. Following the big discoveries of gas in the Groningen province of Holland in 1959, the possibility that similarly favourable geological structures might extend under the North Sea had become fairly apparent, and by the end of 1963, some twenty companies were already nearing completion of seismic, gravity and magnetic surveys in the Southern Basin.[1] All of this was well enough known at the time, as was the fact that to proceed from these preliminary investigations to the more expensive business of drilling involved certain legal difficulties. In terms of the exploitation which alone could ultimately justify such investment, the industry was working in what amounted almost to a legal vacuum, a maritime location outwith territorial waters where no viable system of proprietary rights existed in connection with any mineral resources that might be discovered. In short, one of the most basic prerequisites of the symbiotic relationship between host state and industry had not been fulfilled. It was not just that the defence of property rights was not assured; their very existence was not as yet established. The situation, as one observer described it, was a 'farcical' one, in which 'the companies were carrying out their investigations in the North Sea and yet they could not put down a rig or work a drill because there was no one in the world who could grant them a concession to work the oil or natural gas if they struck it.'[2]

Farcical perhaps, but not quite accurate inasmuch as the power to create a framework within which such concessions could be granted was to hand. For almost twenty years, following the Truman Proclamation of 1945 *vis-à-vis* the US Continental Shelf, states had been claiming varying degrees of sovereignty over the shelves abutting their own and dependent territories.[3] Moreover, in 1958 the First United Nations (UN) Conference on the Law of the Sea had produced, among other things, a Convention on the Continental Shelf which, upon ratification by twenty-two countries, would accord sovereign rights to coastal states in connection with the exploration and exploitation of the natural resources of their shelves. While the Convention, despite its terminology, did not apparently envisage sovereignty to the extent of full *dominion* (ownership of the territory), it certainly did allow for the creation of a legal regime that would establish the requisite property rights in offshore oil and gas.[4] Furthermore, by late 1963, the Convention lacked but two more ratifications to bring it into force.

Some mystery still surrounds the delay countenanced by the Conservative Government of the day over enacting the legislation which would eventually make Britain's the last of the necessary ratifications of the Convention on the Continental Shelf and would set up a legal framework for offshore operations. By its own account, the need for legislation only became clear in 1963, with the completion of seismic surveys and expressions of extreme interest in further developments by the companies, in the absence of which any earlier move would have been a waste of Parliament's time.[5] What is not in doubt, however, is that once the industry (though not the administration) possessed information to justify further exploratory activity, it would not readily concur in political dilatoriness. Indeed, one senior civil servant was later to recall that although the companies would never have pressed ahead in the absence of a legal framework, if there had been any earlier reason to suspect the existence of very large reserves of gas or oil in the North Sea, 'they would have brought it very forcibly to the attention of the Government.'[6] The Minister of Power put the point very clearly at the time:

> we know that many of these companies are interested to go further and secure as soon as possible the right to begin drilling. Surveys, however promising, can do no more than indicate the existence of

geological structures that might contain oil or natural gas: only drilling can determine whether oil or natural gas is actually present. But drilling, particularly under the sea, is a very expensive proposition. . . . At this point the companies have very reasonably turned to the Government and asked that their operations and investments shall have the protection of a proper system of law. This is something which only Parliament can provide.[7]

Whatever the reasons for its previous tardiness, the Government now responded to this request with a degree of legislative alacrity tantamount almost to unseemly haste and, in the minds of some, to a more than slightly cavalier attitude towards proper Parliamentary procedure. Thus, for example, when the Continental Shelf Bill, introduced in the Lords in November 1963, reached the committee stage in that house on 17 December, the Minister of State at the Home Office invited those who were already expressing considerable misgivings – not least about safety, as we shall see – to let the Government have details of any particulars in which the general powers encompassed in the Bill did not seem to cover all the envisaged contingencies.[8] Startled enough that an administration 'normally . . . not afraid to govern in this sort of matter' should now be relying on piecemeal suggestions,[9] the critics were to be staggered when, on the following day, they discovered that the Bill was down for its Third Reading on 19 December. 'Preposterous' was one verdict, while Lord Shackleton wanted to know, not unreasonably, 'how it could be expected that we could come forward with suggestions between nine o'clock last night and tomorrow'. In the end, the affair was explained away as a 'slip-up in the usual channels', with the Minister of State, Lord Derwent, conceding that he had been 'too forthcoming' and his opponents rubbing salt in the wound by acknowledging his helpfulness 'in the extremely limited time that the Government have allowed him'. Thus the proposed enactment proceeded to the Commons without, as Lord Shackleton again expressed it, 'our having done our work on it as thoroughly as we should'.[10]

In the Lower House this momentum was sustained throughout the first three months of 1964, and no one was left in any doubt as to the urgency of legislating. Time and again, the need for haste was emphasized, and short shrift given to those (admittedly, few) who might have wished to proceed at a less ill-considered pace. Probably the most disreputable moment in this not very commendable chapter of British legislative history occurred when the Minister of Power, Mr F. J. Errol, even chided a Standing Committee of the House for the fact that the 'biggest delay' was being occasioned by its own deliberations and, in terms redolent of a toolpusher rather than of a Government Minister, went on to explain that 'Every week counts.'[11] Not suprisingly, and although he hastened to temper his irritation by insisting that he was not complaining, his statement earned him a stern rebuke:

> The Minister was hardly fair in suggesting that we were already holding things up by our discussion of the Bill. To accept that argument would place us under a handicap. If things are being held up, the blame is his and not ours. It is rather more than Parliament can stomach to be told that it must not discuss an important subject, yet that was implicit in what the right hon. Gentleman said.[12]

One obvious explanation for the aura of unconscionable haste surrounding this enactment is that the Conservative Government was anxious to have it on the Statute Book, and on its own terms, before facing an imminent general election. Nor did this possibility escape Labour members, some of whom lost no opportunity to play electioneering games, such as trying to force a Tory administration into a public admission that its proposed legislation would effectively nationalize the resources of the Continental Shelf.[13] But behind all the legislative speed there also, of course, lay other (and familiar) factors. Indeed, virtually the first fly cast to the parliamentary fish when the Bill was introduced in the Lords was tied from benefits to the balance of payments, savings in foreign exchange and reduced dependence on overseas supplies.[14] So too in the Commons, the bait was quickly reeled out in similar guise and the House invited to agree that 'if oil or natural gas could be found in quantity close at hand, and under our own control, it would be a matter of great good fortune.'[15] Any such good fortune, moreover, was not to be subjected to any deferral of gratification. The whole approach was summed up in the words with which the Minister concluded his opening speech on the Bill's Second Reading: 'The main purpose of the Bill, in despite of the many incidental provisions, is to enable the natural resources of the Continental Shelf to be exploited for the benefit of all concerned. I hope that the House will agree that this is highly desirable and urgent. . . .'[16]

To some of these 'incidental' issues – incidental to the Bill's main purpose, perhaps, but certainly not to the welfare of those whose labour would be vital in accomplishing it – we will shortly return. Here, however, it is worth lingering just a little longer over one particularly extreme example of the way in which urgency dominated the proceedings and thereby permeated the legislative atmosphere in which considerations of safety would be discussed. Under Clause 1 of the proposed enactment it was envisaged that the rights over the Continental Shelf available to the UK in international law should be vested in the Crown, and that the areas within which these rights would be exercisable should be defined by Order in Council, creating 'designated areas'.[17] While Labour members were at pains to acknowledge that Government business would soon grind to a halt if every executive act carried out in this manner were to be subject to a parliamentary power of annulment – by negative resolution inside a period of forty days – they nonetheless wondered whether a power of such importance as that proposed by the Bill should not be regarded as one of the instances in which provision for such negative scrutiny would be appropriate. Accordingly, an amendment to this effect was tabled, but the reply was that, 'solely for practical reasons', this was a case in which 'it is necessary for Parliament to forgo what one might almost call a natural right to debate an Order in Council.' As for the 'practical reasons', they were openly admitted – the procedure might cause delay![18] Just how desperate the Government was to press ahead rapidly, not just with legislating but also with exploration, is evident from the following statement, which, not suprisingly, was later to provoke accusations that the issue was being treated as 'far too important a matter to leave to Parliament':[19]

> all orders subject to the negative Resolution procedure have the feature of delay. The point is that the oil interests themselves . . . would be reluctant to embark on costly operations until they know that the licence granted under what to them would be only a temporary order would become a definite one because the period of forty days had passed.
> This is not a matter of months. We know that the oil companies are extremely anxious to get going as quickly as possible. I emphasize the time factor in this matter. It is much easier to work in the North Sea in the summer than in the winter, and it is not a matter of waiting for months before anything happens.[20]

In an atmosphere such as this, it was highly unlikely that any doubts about the adequacy of arrangements for securing the safety of offshore operations would be allowed to impede the rapid provision of the legal framework required for further exploration. Nor were they, even though safety was one of the major misgivings raised by the Bill's critics from the very beginning and even though the doubts which they expressed also exposed some major elements of confusion in the

144 The Growth of a Special Regulatory Framework

proposed enactment's crucial provisions for the application of British law in general. Instead, Government spokesmen floundered from explanation to explanation and from assurance to assurance, clarifying nothing, giving little away and, in the end, doing much to create the legal chaos which has been the hallmark of the North Sea safety regime ever since. In its panic to bring what it ingenuously called 'law and order' to the British Continental Shelf,[21] the Government produced a statute that must stand as a fairly damning indictment of a legislature which, as more than one speaker noted at the time, already had nearly 150 years' experience of safety legislation.[22]

Left to its own devices, the Government's intentions with regard to offshore safety were fairly straightforward. Apart from such matters as protecting installations from the intrusion of shipping and vice versa, dealt with separately under Clauses 2 and 4, the plan was to make use of Section 6 of the Petroleum (Production) Act of 1934, which was now being extended to the Continental Shelf.[23] By the terms of that Section, the Secretary of State was required, before issuing any licences, to make regulations prescribing, amony other things, model clauses which would normally be incorporated in the licence. On land, this procedure already encompassed a system whereby licensees were required to comply with any instructions for securing the safety and health of employees which the Minister of Power might issue from time to time, and the intention now was to apply this 'proved procedure' to the 'new situation'. Moreover, the Institute of Petroleum was already obligingly revising its Code of Safe Practice to take account of offshore developments, and when this review was complete, licensees could be instructed to comply with its recommendations. If this was not sufficient, additional instructions could be issued.[24]

The trouble was, however, that none of these intentions, however laudable, was anywhere stated explicitly in the Bill, in itself perhaps an indication of the importance being accorded to safety in the rush to legislate. Indeed, their existence was only elicited by protracted questioning in both Houses, where, despite protestations to the contrary, the distinct impression was created that an attempt was being made to push the Bill through without any satisfactory answers being given. This impression was modified not a whit by the fact that it was primarily discussion of the safety issue which was curtailed by the 'slip-up in the usual channels' referred to above. Moreover, closer scrutiny revealed that while Section 6 of the Petroleum (Production) Act certainly referred to 'model clauses', it did not specify that these had to say anything whatsoever about the safety, health and welfare of employees.[25] Thus suspicion mounted that the Government's real intention was to accept 'voluntary arrangements' rather than to rely on any explicit legislative provision.[26]

Such suspicions were by no means allayed as a result of the stance adopted by Government spokesmen with regard to an alternative possibility, namely, that the proposed statute either did or should be made to apply appropriate parts of the Factories Act of 1961 to the Continental Shelf.[27] On this subject they got themselves into an almost unbelievable tangle – unbelievable, that is, until it is once again recalled that the whole affair was so precipitate that, initially, an undertaking could not even be given that a map of the Continental Shelf could be made available![28] At first, the excuse was that 'the ordinary law – factory laws we may loosely call them – could not be applied without a very elaborate scheme of inspection', a fair enough point, but one which scarcely augured well for the future of enforcement under the Government's own scheme.[29] Subsequently, however, recourse was had to another argument hinging upon the complex generic provisions of Clause 3, which purported to apply the criminal and civil law of Britian to offshore operations:

> *Application of criminal and civil law*
> 3 – (1) Any act or omission which –
>> (a) takes place on, under or above an installation in a designated area or any waters within five hundred metres of such an installation; and
>> (b) would, if taking place in any part of the United Kingdom, constitute an offence under the law in force in that part,
>
> shall be treated for the purposes of that law as taking place in that part.
>
> (2) Her Majesty, may by Order in Council make provision for the determination, in accordance with the law in force in such part of the United Kingdom as may be specified in the Order, of questions arising out of acts or omissions taking place in a designated area, or in any part of such an area, in connection with the exploration of the sea bed or subsoil or the exploitation of their natural resources, and for conferring jurisdiction with respect to such questions on courts in any part of the United Kingdom.

This Clause, subsequently to become Section 3 of the Act, was to play its full part in causing many a jurisdictional headache in the years ahead.[30] At this stage, however, the question was: did it mean that the Factories Acts would apply to offshore installations? To this the official answer, though technically correct, was not terribly helpful –

yes, inasmuch as any relevant act or omission occuring on an installation could be defined as taking place in a factory.[31] As Lord Derwent succinctly explained to some highly puzzled peers, 'The Acts that the Noble Lord mentioned, such as the Factories Acts, and so on, where they apply, can be made to apply to an installation on the Shelf.'[32] Interestingly, too, this proposition was consistently advanced under Clause 3(1), which was intended to 'export' the criminal law and which, for a time at least, the Government seems to have regarded as so watertight with regard to the Factories Act that it could safely offer to consider any 'legal gap or flaw' that could be pointed out.[33]

Unfortunately, however, things were not quite as straightforward. What if installations were not factories? What if they necessitated special regulations beyond the scope of the Factories Acts? Furthermore, Clause 3(2), which the Government categorically insisted was designed to deal with the civil rather than the criminal law, carried with it the highly confusing possibility that it might be used to extend enactments containing criminal sanctions. Nor was this mere speculation on the basis of careful textual scrutiny of the Clause itself. Further on in the Bill, its powers were invoked to make specific provision for the application of some such statutes, including the Radioactive Substances Act of 1960 and the Wireless Telegraphy Act of 1949.[34] Hence arose the question: if Government was prepared to make the possible application of these statutes quite explicit, why should it not do the same with regard to safety legislation?[35] The excuse that these additional provisions were not really necessary in the light of the general enabling powers contained in Clause 3(2), and that they had only been included in order to 'make assurance doubly sure' – in terms of Article 25 of the Convention on the High Seas on the one hand, and vis-à-vis the anxieties of the Post Office on the other – never really sounded anything more than lame.[36]

What it all came down to in the end was that the Government was not, if it could possibly avoid it, going to commit itself to clear and probably time-consuming revisions which would make provision for the safety, health and welfare of offshore employees. Thus the attempts which were made in both Houses to secure amendments that would make the application of the Factories Acts quite explicit and would confer powers for the promulgation of safety regulations quite distinct from licence terms were stiffly resisted. On the first score, it was finally and openly conceded that very few of the operations likely to be conducted on installations would, in fact, be covered by the relevant Acts,[37] a confession which could only make the earlier play on Clause

3(1) look like sleight of hand; on the second, it was contended that the extensive amendments which would be required in order to cover such things as the appointment of inspectors, their powers and the procedure for making regulations would simply be inappropriate.[38] Against a background of increasingly forthright comments about affording labour the same degree of protection as capital,[39] the only concession to be granted was that the Minister would explicitly commit himself to doing what he had all along intended. Accordingly, an additional subsection was inserted into the Clause dealing with licensing:

> (4) Model clauses prescribed under section 6 of the Petroleum (Production) Act 1934 as applied by the preceding subsection shall include provision for the safety, health and welfare of persons employed on operations undertaken under the authority of any licence granted under that Act as so applied.[40]

In terms of its primary objective, the creation of a legal framework to facilitate further offshore developments, the Continental Shelf Act of 1964 did its job well enough.[41] As we have already seen, it vested all exercisable rights with respect to the natural resources of the sea bed and subsoil (with the exception of coal) in the Crown, thereby establishing rights which, while possibly not amounting to full ownership, were effectively proprietary.[42] Similarly, it extended the land-based licensing system to the Shelf and thus permitted the business of drilling to get under way with the licensing rounds of 1964 and 1965, the first carried out by the Conservative Government in its dying days and the second by its Labour successor. Interestingly, the reliance which was placed upon the provisions of the 1934 Petroleum (Production) Act in this context did not necessarily entail the granting of licences or concessions to private operators, since the enactment is merely permissive in this respect, and the Government may therefore exploit its own resources. As Daintith has pointed out, however, there is no evidence that any other mode of legal organization was contemplated by the Conservative Government. In terms of the arguments advanced earlier in this book (and implicit throughout the present discussion), his explanation is significant:

> Good economic reasons can be found for this basic decision: rapid exploitation of the resources was desired, mainly for balance-of-payments reasons, and the requisite expertise and capability did not exist in the United Kingdom public sector (nor, indeed, if the first aim were to be vigorously pursued, in the private sector either). Hence it

seemed natural, particularly to the Conservative Government in power in those years, to continue, through the concession system, to make the maximum call on the resources of the worldwide private oil industry.[43]

From what has already been said, it will be apparent that the 1964 Act created a somewhat extraordinary situation with regard to safety. Although at least one earlier researcher has found evidence of an attempt to invoke the Factories Act (by the Department of Employment),[44] the generally accepted position, endorsed by the author in question, is that offshore installations did not come within the definition of a 'factory', as given in the relevant legislation. In consequence, the principal legal protection with regard to offshore occupational safety lay in the instructions issued under model clauses which comprised part of the licence itself. Moreover, while there has been some debate as to whether these model clauses and attendant instructions were of a contractual or a regulatory character,[45] there is no doubt about the nature of the sanctions which were available in the event of non-compliance. The Minister could either effect entry and cause any necessary rectifications to be carried out at the licensee's expense, or he could fall back on the contractual nature of the licensing arrangement and revoke the licence.[46] Despite the efficacy claimed for it at the time, the first option always presented, at best, a rather improbable scenario; the second, like other draconian measures to be discussed later,[47] might certainly be effective if (and it is a very big 'if') the power was used.

The regulatory implications of exercising control over safety by these methods became tragically apparent within some eighteen months of the Act's passage when, on 27 December 1965, the 'jack-up' rig Sea Gem collapsed and sank with the loss of thirteen lives. One incidental, though by no means unimportant, flaw quickly became evident inasmuch as the Tribunal which the Minister of Power set up to investigate the circumstances of the collapse was forced to operate without statutory authority and without power to compel the attendance of witnesses or to administer oaths.[48] With the generous co-operation of all concerned, however, the Tribunal was able to carry out its duties satisfactorily, and the report was completed by the end of July 1967. Failure of tie bars on the port-side jacks, probably initiated by the brittle propensities of the steel at the ambient temperature on the day in question, was found to be the most likely cause of the disaster.[49]

But the Tribunal had also been instructed to look into the safety procedures applicable to the installation, and on this score the picture which emerged was far from reassuring. BP Petroleum Development Limited had been granted a licence on 17 September 1964, incorporated into which had been Model Clause 18, as set out in Schedule 2 of the Petroleum (Production) (Continental Shelf and Territorial Sea) Regulations of the same year. Under this clause, the licensee was required to 'comply with any instructions from time to time given by the Minister in writing for securing the safety, health and welfare of persons employed in or about the licensed area'. In November, the Minister had indeed issued such instructions, and in terms typical of the most important instructions given to licensees at that time, had told the company that

> until further notice all operations in or about the licensed area to which the licence relates shall be carried out in accordance with such provisions of Part VIII of the Institute of Petroleum Model Code of Safe Practice in the Petroleum Industry issued in October, 1964, as relate to the safety, health and welfare of persons employed.[50]

According to the report, this Code had not been adhered to in 'several important particulars' – not least inasmuch as no escape drills, much less the 'frequent' ones which it recommended, had taken place.[51] Although the Tribunal was careful to avoid any suggestion that this or any of the other important departures from the Code had actually contributed directly to the occurrence of casualties, it nonetheless went out of its way to underline the shortcomings of a system which relied upon this Code for the substance of safety provisions and upon the contractual nature of the licence for their enforcement. For one thing, the report observed, the authors of the Code themselves emphasized that it dealt only with recommendations and not necessarily with 'anything issued in the form of regulations or instructions by the appropriate national authorities'. Moreover, its structure and layout underlined the importance of this reservation, since they were 'not apt to make it clear and authoritative as a piece of quasi-legislation'.[52] As for sanctions, the Tribunal seems to have regarded the Minister's power to take unilateral action as not even worth discussing, but spotted the other main regulatory problem right from the beginning. 'It is perhaps as well to make it clear this early', said the introduction, 'that the only sanction for ensuring the proper operation of the safety procedures is the revocation of the licence.'[53] Accordingly, the three-man inquiry went on to include in its list of recommendations a crucial proposal for a *statutory* code supported by 'credible sanctions'.[54]

In the light of such firm conclusions, it might not have seemed unreasonable to expect that some acceleration would now take place in the pace of regulatory provision. However, while the Labour Government in power from 1964 onwards proceeded with two further licensing rounds which maintained rapid development as the number one priority, it demonstrated nothing like the same sense of urgency over implementing the recommendations of the Sea Gem report. Certainly, further instructions taking the report's conclusions into account were issued, but they were still promulgated under the old system and remained subject to the same limitations.[55] As far as statutory controls are concerned, however, the intention to legislate was not announced until 1969, and even then the Government could not find the requisite time in its parliamentary timetable before leaving office in 1970. The contrast (or possibly the underlying similarity) with 1964, when the Continental Shelf Act had 'only just scraped into a busy parliamentary session, thanks to the dropping out of another Government Bill',[56] is pronounced indeed. It would not perhaps be too unkind to take the legislative alacrity of the Conservatives in 1964 and the subsequent tardiness of their Labour successors as indexing something akin to bipartisan neglect with regard to safety.

From Regulatory Inertia to Statutory Regulation

Five years after the Sea Gem collapsed, and some two and a half years after the report on the circumstances of that collapse was published, Parliament finally got round to the task of making clear statutory provision for the safety of offshore operations. By then, oil as well as gas had been discovered in the North Sea; a new Conservative Government was in power; at least fourteen more workers had been killed since 1965;[57] and three other rigs had followed the Sea Gem, though fortunately not with the same disastrous consequences in terms of loss of life.[58] As one MP put it, 'The Bill should have been on the Statute Book long ago.'[59] In view of the time taken over putting it there, an unusually large measure of the smugness which distinguishes so many official pronouncements on Britian's record in the field of safety legislation must surely have been required for the new Under-Secretary of State for Trade and Industry to put the Bill forward in historically creditable terms:

The whole House is excited by the prospects of the developing industries in oceanology, and we must therefore be ready to allow all these new developments to take place, but to keep alive what has been a typically British tradition – that we are ahead with our safety regulations, not in any sense to cramp expertise and initiative, but in order to ensure that those who take these initiatives are adequately protected in so far as Parliament can protect them by safety legislation. This tradition goes right back to the nineteenth century, and the House should be justly jealous to keep it up to date.[60]

Historical accolades apart, this statement conveys a fairly accurate picture of the atmosphere in which the Mineral Workings (Offshore Installations) Bill was discussed during the first half of 1971. Parliament must now legislate, not least because the inadequacies of the present system were apparent; but in doing so, it must pay due attention to the industry's technological uniqueness and progress. As Earl Ferrers explained in the Lords, 'This is a new and developing industry, and the Government have been particularly anxious to adopt a flexible attitude towards the problems of it.'[61] Nicholas Ridley, Under-Secretary of State at the Department of Trade and Industry, which had now absorbed the old Ministry of Power, took a very similar line in the Commons. Any archaic suggestions that regulations should be enshrined in the Act itself rather than being promulgated under enabling provisions was not to be entertained 'because, with this rapidly changing industry, and its advancing techniques and technology, it will be necessary to bring in new regulations and to change them frequently in response to development'.[62] Reasonable enough in its own terms, this argument did not completely mollify those who seem to have been sensing here that the proposed legislation was going too far down the road towards reducing the regulation of safety to a reactive role in relation to technological advances. Such fears were expressed particularly strongly, for example, over the extensive powers of exemption envisaged in the Bill, powers which some members felt might be making safety a hostage to technological fortune. On this, as on other similar misgivings, however, the answer was the same: there must be flexibility 'because we do not know what is going to happen in the future'.[63]

Just as the legislative intention here clearly did not extend to the creation of a regulatory system which might hamper the industry's spontaneous technological progress, so too great pains were taken to avoid offending its finer susceptibilities. Time and again, for example, and in terms evocative of those nineteenth-century legislators who always insisted that the vast majority of employers did not really require the statutory regulation about to be imposed on them, it was stressed that the industry had always been very safety-conscious and that it was now being very co-operative over the proposed legislation.[64] Similarly, Government spokesmen went out of their way to provide assurances that although legislation was now on the way, it was not intended to be oppressive. To be sure, 'the full force of the law' would be brought down upon 'persistent and deliberate infringement', but the overall plan was to proceed by 'benevolent enforcement . . . generally advisory in nature'.[65] As for regulations, some of them would have to be delayed to allow for consultations, which would 'ensure that we carry the industry with us in believing that [they] are fair, reasonable, effective and practicable'.[66] Even the seemingly trivial question of the title to be given to the person who should be placed unequivocally in charge of every installation, one of the main recommendations of the Sea Gem Tribunal, could not be decided without a remarkable degree of deference being shown to the industry's opinion. When Labour Members objected to what they saw as the hierarchical Merchant Navy connotations of the term 'master', Mr Ridley agreed to reconsider. But in doing so, he made it clear not only that the industry's views were important in this matter, but even that they should take precedence over the opinions of either Government or Opposition.[67] Once again, it took a jealous parliamentarian to suggest that the powers of Commons Committees should not so readily be usurped, even in matters of terminological dispute:

> Although the industry and the Government have had private negotiations, there is nothing to prevent this Committee from saying, as other Committees have said . . . that, though the Government have taken soundings, and though the industry has given its opinion, we as legislators will take our decision. No matter how important the views of the industry and no matter what regard should be paid to them, there will be occasions when we, having listened to the industry's views, will take our own decision because we think that it is right.[68]

That the Mineral Workings (Offshore Installations) Act of 1971[69] should have been passed in an atmosphere of such apparent deference is not suprising when it is viewed against the broader background of the relationships and developments discussed above. For one thing, although the Continental Shelf Act had fulfilled one vital part of the state's symbiotic obligation by creating a system of proprietary rights for the North Sea, this did not mean that the relationship had there-after become a one-way affair. With exploration moving into the more northerly and much more expensive areas, there was a continuing 'necessity' for Governments to offer effective guarantees that the rules of the game would not be radically altered in such a way as to compromise cost profiles or timescales, and this is arguably what heavy-handed safety legislation might well have appeared to be doing. From a somewhat later point in the history of North Sea development, for example, Noreng estimates that some 20 per cent of the $1716 million cost difference between the Brent D project and the compar-able Statfjord B is accounted for by the stricter safety regulations applying in the Norwegian sector.[70] Nor should it be forgotten that the period during which this Bill was under discussion was one when the Government, in response to an apparent fall-off in activity and an increase of some £200 million in the foreign-exchange cost of oil imports, was preparing to embark upon a fourth and highly attractive round of licensing designed to stimulate a rapid up-turn in North Sea activity.[71] Thus to have threatened stringent enforcement of inflexible regulations might have detracted from the allure of the new blocks to be put on offer and thereby run counter to other declared aims of official policy.

This Act, which was passed in July 1971, was to become the main legal instrument for controlling the safety of offshore operations in the years ahead. Tailor-made, so to speak, it gave the offshore oil industry its own safety legislation, and this at a time when, ironically enough, an official enquiry into health and safety at work in general was just reaching the half-way point in its deliberations which would condemn the proliferation of such piecemeal provision as one of the main deficiencies of the prevailing system.[72] While this issue was not raised at the time, the Act (and its associated administrative infrastructure) was in a sense almost an anachronism even before it came into effect. Although we shall return to this side of the matter and to some of the new law's detailed requirements at a later stage, for purposes of the present discussion, its main provisions may be summarized as follows:

1. The Act was to apply to exploration for, and exploitation of, mineral resources within territorial waters or areas of the Conti-nental Shelf designated under Section 1(7) of the Continental Shelf Act 1964. The operations in question had to be taking place on a floating or other installation unconnected to dry land by any permanent structure. (Section 1)

2. Installation managers (the term finally agreed) were to be appointed, and their responsibilities were defined, as were those of the owners of both installations and concessions. (Section 2, 4, 5 and passim)
3. The Secretary of State was empowered to make regulations covering, among other things:
   (a) the registration of installations (Section 2);
   (b) the construction and survey of installations (Section 3);
   (c) the safety, health and welfare of those aboard, whether employed or not (Section 6);
   (d) the safety of installations and the prevention of accidents on or near them in connection with transportation, with vessel and aircraft movements and with any operation on or near the installation, whether in or out of the water (Section 6);
   (e) the appointment of managers (Section 4);
   (f) the appointment of inspectors and their powers (Section 6 and Schedule);
   (g) the keeping of records and the reporting of accidents (Section 6 and Schedule).
4. Constables were to have the same powers, protection and privileges that they have on land. (Section 9)
5. Except where specially restricted, offences were to be punished by a fine not exceeding £400 on summary conviction and imprisonment for a term not exceeding two years (with the possible addition of a fine) upon conviction on indictment).

From this list it is fairly evident that the subsequent pace at which legal control would be asserted over the safety of offshore operations was going to depend very heavily upon how quickly the Secretary of State made use of his extensive powers to frame regulations. In this context, however, he may well have had cause to rue the words of his Under-Secretary, who had rashly stated that although 'it may take a little longer to get all the regulations completed', the intention was 'to get the first regulations complete and published within six months'.[73] In the event, it was nearly a year before the first two sets of regulations, those relating to registration and managers, were laid before Parliament.[74] A further six months elapsed before the Offshore Installations (Logbooks and Registration of Death) Regulations came into effect in November 1972,[75] and thereafter another whole year went by before regulations covering inspectors and the reporting of casualties were made.[76] Once again, the contrast between the speed of developments in the North Sea, where activity had picked up dramatically in the wake of the fourth licensing round, and the laggardly pace of safety regulations was becoming apparent. To hoist the authorities with their own historical petard somewhat, it is not perhaps unfair to point out that nearly 150 years earlier, Britain's first four factory inspectors took only three years to devise a completely novel code of regulations dealing with a much more complex set of administrative matters and covering no less than 3000 textile factories with a workforce of nearly 200,000![77]

Within a month of the Offshore Installations (Inspectors and Casualties) Regulations' coming into effect in December 1973, the gap was dramatically highlighted when the capsize of the Transocean 3 – an accident which some experts claim should have taught lessons that might have averted the full horror of the Alexander Keilland disaster – provoked a further parliamentary debate during the dying weeks of the Conservative Government. Pressed upon the question of how his Department, now yet again hived off into a separate Department of Energy, was coming along with the business of making regulations, Mr Peter Emery was able to claim, justly, that four sets had already been made, and no less fairly, that two others were in an advanced state. As we have already seen, however, the regulations in operation at this stage were largely concerned with the administrative infrastructure rather than with the more detailed substance of safety requirements, and there was therefore no dodging the damaging admission that 'the present method of enforcing safety requirements is by non-statutory conditions attached to petroleum exploration and production licences granted by the Department.'[78] In other words, and although he did not put it quite like this, six and a half years after the replacement of the licence-linked system had been recommended, it was still in use as the main instrument of control over offshore safety. Even the two and a half years which had now elapsed since the requisite powers had been established struck some as 'far too long a time to wait', frontiers of technology and understandable problems of engineering design notwithstanding.[79]

At this time, there were certainly some signs that a more jaundiced reaction was beginning to emerge with regard to the regulatory record of successive Governments. Accusations of profit being put before safety, of possible corner-cutting, of a wartime attitude which treated men and materials as expendable and of a general lack of planning being allowed to carry over into the realm of safety were all advanced during this brief debate. Nor were they effectively refuted by counter-charges of ignorance, paeans of praise for the effectiveness of an already discredited system and bland assurances that the new Department of Energy was 'completely and utterly sold on the need for adequate safety provisions and regulations'.[80] Moreover, some critics were now beginning to sense something of the connection between the political economy of the North Sea and the adequacy of the legal regime governing safety. Margo MacDonald (a leading Scottish Nationalist Member), for example, linked the whole question of safety regulations to the forces underpinning rapid development and to the particular kind of Government/industry relationship which they had spawned:

> I would suggest that the lack of safety regulations is symptomatic of the attitude of panic by the Government towards the oil industry generally. Since I came to this House all I have heard about has been the serious economic situation. All I have heard is that the way out is for the Government to get their hands on the oil as quickly as possible, to take the oil and the revenues. That is why the Government are prepared to put exploitation of the oil before safety. The Department of Trade and Industry and now the Department of Energy appear to be permitting proper considerations of safety for those directly and indirectly involved to be bypassed in the haste to get hold of the oil and the revenues. . . . We all know that the Government are afraid to deal more firmly with the oil companies because they have the mistaken idea that the companies may pull out of the North Sea. . . . The longer the oil remains under the sea, the more valuable it becomes. We have plenty of time to make sure that we are geared up technologically, socially and physically to take advantage of the industry and make it work for us. . . .[81]

As we saw in the previous chapter, the advent of a Labour Government in February 1974 brought a belated start to the business of tightening up the terms on which the Continental Shelf would be exploited, even if this did not betoken any dramatic change of heart on the question of rapid development in the short term. So too with safety: the ensuing period saw lost gound being made up on the promulgation of special regulations for the offshore industry, albeit very late in the day. Thus, and although the new adminstration could scarcely claim any of the credit, the glaring omission revealed by the Sea Gem Tribunal's lack of statutory authority (nearly seven years earlier) was finally rectified by the Offshore Installations (Public Inquiries) Regulations, which came into effect in April 1974.[82] A month later, the Offshore Installations (Construction and Survey) Regulations became effective and stipulated that, after the end of August 1975, all installations must be in possession of a valid Certificate of Fitness issued by a Certifying Authority appointed by the Secretary of State.[83] The matters to which the relevant surveys should extend were specified in a schedule, and in June 1974 the first five Authorities were appointed – the American Bureau of Shipping, Bureau Veritas, Det Norske Veritas, Germanischer Lloyd and Lloyds Register of Shipping. All five, it may be noted, were ship-classification societies, and the adequacy of their experience out-with the field of shipping was to be the subject of some concern both at the time and subsequently.[84] Nor will the reader who recalls our earlier description of the pace of offshore development fail to notice that, in the event, the date set for North Sea installations having to be certified as fit for their designated function would turn out to be several months after the first of Britain's oil had actually started to come ashore.[85] While the allowance of a period of grace between the promulgation and the implementation of the regulations is understandable enough in view of the need for familiarization, necessary modifications and so on, the fact that North Sea oil production, with its long lead-in period in terms of exploration and development, should have started before such a basic safety requirement was in operation possibly speaks for itself.

In the meantime, however, a start was also being made on some other long overdue regulatory matters. Rules governing the conduct of diving operations from installations were promulgated in the middle of 1974 and came into effect at the beginning of the following year.[86] In late 1975, the statutory powers were assumed for the making of regulations to cover the safety and inspection of pipelines and pipeline works, thereby plugging a gap which the Secretary of State, though not, surely, anyone familiar with the previous history of safety legislation in this field, found 'astounding'.[87] These powers were subsequently used to make the Submarine Pipelines (Diving Operations) Regulations 1976 and the Submarine Pipelines (Inspectors, etc.) Regulations which came into effect in June 1977.[88]

With reference to installations themselves, this particular part of the regulatory story was completed by four sets of regulations which appeared between 1976 and 1978, or five to seven years after the enabling legislation had been enacted and some nine to eleven years after the need for statutory controls had been pointed out. Nor were they concerned with matters so peripheral that the lateness of their emergence can simply be put down to the execution of a residual tidying-up exercise after the main issues had been dealt with. The Offshore Installations (Operational Safety, Health and Welfare) Regulations of 1976, for example, introduced a wide range of extremely important provisions, as will be apparent from the following list of subjects covered:[89]

1.   designation of hazardous areas;
2.   a system of work permits for use of potential ignition sources and for work in inadequately ventilated areas;
3.   the safe keeping of dangerous substances;
4.   the examination, testing and maintenance of all parts of the installation and its equipment, including lifting gear;
5.   written instructions specifying practices to be followed in connection with drilling production, electrical and mechanical equipment, personal and installation safety procedures;
6.   the provision of sound, safe and suitable equipment and of protective clothing and equipment;
7.   safe place of work and means of access;
8.   radio communication equipment and operation, and means of ascertaining prevailing environmental conditions;
9.   safety of helicopter operations and control of landing by sea or air;
10. provision of water, food and medical facilities;
11. a ban on the employment of anyone under 18 and the keeping of records relating to hours worked.

Though not so wide-ranging, the importance of the remaining three sets of regulations is also apparent. The Offshore Installations (Emergency Procedures) Regulations 1976 required the provision of an emergency procedure manual and muster list, the holding of drills and the provision of a safety vessel. Lifebuoys, lifejackets, survival craft and general alarm and public-address systems were made statutory requirements by the Offshore Installations (Life-Saving Appliances) Regulations of 1977. Finally, and after what appear to have been particularly protracted negotiations with the industry, regulations pertaining to the provision of fire-fighting equipment were made in 1978, with certification requirements to become operative in April 1980.[90] While all the technological and other difficulties must be fully acknowledged, a salutory reminder of how development outran safety regulation in this field is provided by the fact that only in the fourth month of the year in which Britain was meant to become self-sufficient in oil did it become an offence to allow anyone on to an offshore installation unless the fire-fighting equipment and plans had been officially examined and certified within the preceding two years.

The Growth of a Special Administrative Framework

Thus far, the regulatory framework pertaining to the safety of offshore operations has been discussed in terms of the evolution of rules, principally those contained in, or made under, the auspices of the Mineral Workings (Offshore Installations) Act. Important as this aspect of the North Sea's regulatory history may be, however, it does not provide a complete picture. As we shall see in a subsequent chapter, for example, although the enactments and regulations dealt with up to now certainly constitute the most important body of rules concerning offshore safety, they do not exhaust the statutory repertoire.[91] Equally, and even within the less diffuse context of the specialized legislation discussed thus far, it is obvious that substantive rules do not, on their own, comprise the entire regulatory framework. Unless rule making is to be nothing more than a paper exercise, a gesture of little more than some possible symbolic significance, it must also involve the creation or adaptation of administrative structures with a view to ensuring that, at least in some measure, the rules are implemented.

In the present instance, such administrative developments are all the more important because, as we have seen, so much of the actual rule-making process was delegated under the enabling provisions of the relevant legislation. As a result, consideration of how the regulations themselves were formulated cannot be divorced from the question of the administrative structures which emerged to undertake this task. Nor can this issue be dealt with simply in terms of the administrative arrangements which were made for the specific issue of safety regulation; such arrangements formed only one part of a broader pattern of administrative development relating to offshore operations as a whole, and it is only within this wider setting that the growth of a specialized regulatory machine can be understood.

On this more general plane, the pattern of offshore administrative development has recently been examined by Øystein Noreng in his book on the strategies adopted by the British and Norwegian Governments with respect to the North Sea oil industry as a whole.[92] According to this author, two possible administrative frameworks were available. On the one hand, he suggests, use could be made of 'a horizontal structure consisting of hierarchical layers of Government agencies with generalized functions', already existing agencies which would simply extend their domain of responsibility sideways to incorporate relevant aspects of offshore operations. On the other hand, resort could be had to what he describes as 'a vertical structure consisting of functionally defined sectors of Government, with agencies having specialized functions'. Having outlined the relative advantages of each strategy – in terms of broad 'macro-effectiveness' and more focused 'micro-effectiveness' respectively – Noreng goes on to describe how both countries moved towards vesting the main responsibility for North Sea operations in primary structures which are vertically organized. While secondary, horizontal structures continue to be involved in offshore affairs, it is his contention that they have little influence as against the dominant power of the vertical structure in either country. Significantly, the example which is chosen for the purpose of illustrating this distribution of power with respect to Britain is the fact that responsibility for preventing accidents rests with the Department of Energy rather than with 'Government agencies for social affairs or environmental protection'.[93]

Professor Noreng is in no doubt as to why this particular pattern of offshore administration should have developed. Certainly, he concedes, the vertically integrated approach does bring problems, but it nonetheless offered the overriding advantage of enhanced rationality in decision making with regard to a highly important and specialized area of the economy: for one thing, it increased the 'detailed insight' available to a Government which, he agrees, 'had inferior information resources compared to the industry';[94] more generally, its primacy gave decision makers the power to pursue policies of optimal efficiency over a functionally specific field. For Noreng, recourse to vertical structuring is almost the *sine qua non* of a rational policy with regard to North Sea oil:

> In both countries the main responsibility and powers are embodied in the vertically organized primary structures. The powers give the organizations of the primary structure the freedom of action required to make rational decisions, weigh preferences and have the necessary insight and knowledge to maximize expected utility. This is in many ways also a prerequisite for the development and implementation of a rational oil policy. However, the powers also enable the organization of the primary structure to dominate oil policy, and in particular in most matters related to oil to dominate the organization of the secondary structure.[95]

This general analysis of North Sea administration is not without its faults. What seems to be the fairly unequivocal relegation of the

Treasury to secondary status, for example, is both highly questionable in terms of policy constraints and arguably something over which the author adopts a somewhat contradictory position when he comes to discuss the role of that body in relation to BNOC.[96] Equally, Noreng's mode of bifurcating micro- and macro-effectiveness leaves something to be desired, inasmuch as the latter is merely equated with a vague notion of 'steering at a social and political level', a notion which at times deteriorates into nothing more than the unexplicated idea of the capacity of 'society's values and preferences' to impose themselves on politicians. Conversely, his identification of micro-effectiveness with 'steering at a detailed level, according to specific and technical criteria' seems to neglect the extent to which, and the processes whereby, political pressure does find expression at this level.[97] Not least, I would argue that as far as the British sector of the North Sea is concerned, he underestimates the impact which the underlying concern for speed had upon the perpetuation of a vertical administrative structure incorporating responsibility for offshore safety.

All this said, however, there is no doubt that the history of administrative developments pertaining to the British sector of the North Sea broadly conforms to the pattern outlined by Noreng. Initially, as has already been mentioned in passing, the onus of handling offshore affairs, including safety, fell upon the Ministry of Power, within which a Petroleum Division had already been established for some time. Headed by Angus Beckett, whose role in shaping early Continental Shelf policy was subsequently to come in for some heavy criticism,[98] this Division saw the appointment of the first two inspectors in 1966 and 1968 respectively.[99] In the following year, however, the Division followed its parent body into the Ministry of Technology as a result of a reorganization involving 'two main areas where the responsibilities affecting a number of Departments are concentrated under a senior Minister, following the principle which has already been applied in the fields of defence, overseas affairs and social security'.[100] Thus the Petroleum Division became integrated with a wider administrative enterprise, which also assumed responsibility for many of the industrial functions of the old Board of Trade. The main development in connection with offshore safety during this period was the establishment of a Petroleum Production Inspectorate as part of the Division.

But the Division's sojourn in the Ministry of Technology was to be brief. In 1970, the Conservative Government announced a major restructuring of departmental responsibilities according to 'functional principles' and designed to avoid 'the difficulty of co-ordination which organization by area or by client group would involve', a difficulty which it acknowledged as one of the obstacles to 'improved policy formulation and decision making' in an increasingly complex society.[101] Accordingly, most matters pertaining to offshore gas and oil now became the business of a unified Department of Trade and Industry with a vast range of responsibilities, running from consumer affairs and overseas trade to industrial development, civil aviation, aerospace research and, of course, petroleum. Government attempts to break with client-based organizations notwithstanding, however, the growing preoccupation with oil and energy affairs from 1970 onwards was reflected in the Department's internal structure. Thus by 1973, and

[UNDER-SECRETARY]

BRANCH 1	BRANCH 2	BRANCH 3
UK licensing for petroleum exploration and production, general policy and legislation.	UK licensing for petroleum exploration and production, operational control, safety, health and welfare of workpeople, research, royalties, pollution matters	UK licensing for petroleum exploration and production, technical services, inspection of operations, assessment of discoveries and prospects

Figure 3
Petroleum Production Division (September 1973)
Source: Civil Service Yearbook, HMSO 1974–9. Crown copyright.

although they comprised part of a wider Industry Group, two of the Department's divisions were dealing exclusively with the oil industry, the Oil Policy Division with two branches (Home and Overseas) and the Petroleum Production Division with three. In addition, the Offshore Supplies Office had been belatedly established as a separate unit within the Department's Industrial Development Executive, while oil and gas technology had been developed to a separate section within Branch 2 of the Energy Technology Division. With the exception of pipelines, safety was primarily the responsibility of the Petroleum Production Division (see figure 3).[102]

Once again, however, these arrangements were to be short-lived. In the fraught winter of 1973–4, Edward Heath found himself under increasing pressure from opponents and supporters alike to hive off energy responsibilities from the broader bailiwick of the Department of Trade and Industry. Faced with a grave economic situation, an 'energy crisis' and looming discontent among the miners, he pondered this advice during the Christmas recess and eventually came to 'a firm decision that all energy issues should be concentrated under one of his most senior Ministers and closest colleagues', Lord Carrington.[103] Accordingly on 8 January 1974, the establishment of a separate Department of Energy was announced. Thus despite a period of what might be called administrative vagrancy (occasioned more by the attempts of successive Governments to co-ordinate their own decision-making processes than by any particular concern over energy policy), responsibility for North Sea operations came to rest upon the shoulders of an agency for which energy production was the primary concern.

While the emergence of this new Department did not put offshore oil and gas completely out on their own, administratively speaking, it did herald a new era in which energy issues would be accorded a heightened degree of salience in policy making. As the *Financial Times* remarked on the occasion of the Department's first anniversary. 'Its formation marked the first really concentrated effort since the War to develop and oversee energy policy as a major issue in its own right.'[104] Moreover, within this brief – which still included responsibility for coal, electricity, gas and nuclear power – the growing importance attached to the resources of the Continental Shelf continued to make itself felt in both numerical and organizational terms. By the end of 1977, for example, over a quarter of the Department's complement of staff was directly involved with the oil industry in its various aspects.[105] On the organizational front, a Continental Shelf Policy Division was created in 1975, subsequently to become the more provocatively entitled Continental Shelf (Participation) Division before being largely absorbed into the Petroleum Production Division in 1978. Similarly, in 1976 a specifically designated Offshore Technology Unit was added to the Chief Scientist Group. Of more direct relevance to the immediate exigencies of safety, the following year saw the emergence of the Petroleum Engineering Division (PED), which took over responsibility for most operational and safety matters. As of November 1978, a crucial turning-point in the administrative history of the North Sea, the relevant part of the Department's structure was as set out overleaf.[106]

From the safety point of view, the salient feature of the developments which have been chronicled is the way in which, throughout, responsibility for the bulk of this aspect of offshore operations was

Gas, petroleum production, petroleum engineering
and the offshore supplies office
(Deputy-Secretary)

Gas Division (Under-Secretary)	Three branches.

*Petroleum Production
Division*
(Under-Secretary)

Branch 1	Questions of general North Sea policy; Sullom Voe; assessment and collection of landward and seaward royalties.
Branch II	UK petroleum exploration and production licensing; development of cross-boundary-line fields.
Branch III	British National Oil Corporation.
Branch IV	Oilfield participation policy and participation and financing negotiations with oil companies; policy questions on offshore oil and pollution and Law of the Sea matters; delineation and designation of UK Continental Shelf.

*Petroleum Engineering
Division*
(Director)

Branch I	Administrative organization of development programme approvals; limitation notices, gas flaring, pipeline authorizations, pollution, liaison with other Government Departments; issue of regulations under Mineral Workings (Offshore Installations) Act 1971 and Petroleum and Submarine Pipelines Act 1975; protection of offshore installations and liaison with others on offshore safety, etc.
Branch II	Assessment of petrolem discoveries and prospects; conservation; development plan approvals; petroleum revenue tax field determinations; reservoir evaluation

	and unitization; monitoring of field behaviour; well records; publication of records.
Branch III	Operations and safety; inspectorial advisory functions in day-to-day operations; safety inspections; accident investigations, drilling and abandonment approvals; measurement of oil and gas production; liaison with Certifying Authorities on safety of design and construction of offshore installations.
Branch IV	Research and development related to safety of offshore installations and the exploration for, and exploitation of, petroleum resources of the UK Continental Shelf.
Diving Inspectorate	Inspection of Continental Shelf diving operations; liaison with other Government Departments on diving operations; accident investigations.
Pipelines Inspectorate	Appraisal of technical submissions dealing with land and submarine pipelines; compilation of safety notices or technical annexes as appropriate; administration of the Pipelines Act 1962; agent for HSE in matters of pipeline safety; inspection of pipelines and associated apparatus as necessary.
Offshore Supplies Office	Four branches.

Source: Civil Service Yearbook, HMSO 1979. Crown copyright.

indeed kept firmly located within a vertical structure which also handled such matters as licensing and general North Sea policy. Not only was administrative organization vertical, in the sense that these diffuse functions were subsumed under single Departments – in the oil era, the Department of Trade and Industry and then the Department of Energy – but also in the sense that within those Departments the officials responsible for safety were answerable to superiors whose brief included other, and not necessarily compatible, objectives. Apparent in the Department of Trade and Industry structure shown in Figure 3, this pattern was carried over into energy, despite some increased division of labour within the relevant segment of that Department. Thus throughout the period which saw the promulgation of the most important sets of regulations governing offshore safety, both this issue and the question of enforcement were placed in embarrassingly close administrative propinquity to decision-making concerned with the implementation of politically determined policies governing such matters as the pace of North Sea development.

That this should have been the case during the early part of the period which saw oil become the focus of offshore attention is in no way surprising. Right up to the mid-1970s it was fairly commonplace for 'sponsoring' Departments, initiated during the Second World War, to look after the safety side of their industrial protegés. From 1974 onwards, however, such conflation of functions within one Department became increasingly anachronistic, as horizontal arrangements for the administration of safety became the dominant pattern across industry as a whole.[107] Thus it is significant that a different pattern should have been perpetuated in the case of North Sea oil. As we shall see in the next chapter, furthermore, the administrative and political challenge which this arrangement provoked during the closing years of the decade was never completely successful and, indeed, was substantially rebuffed in the end. To repeat a point that was made at the outset, not the least of the ways in which North Sea oil was to provide a cameo for important sociological issues was through the embittering effect that it would have on the internal politics of the British state.

Here, however, the point to be underlined is that for some years after 1974, safety continued to be combined with other functions inside a Department which now had rapid development not only as one objective, but as part of its very *raison d'être*. To many politicians at the beginning of that year, for example, oil rather than the miners' dispute lay at the heart of the crisis,[108] and organizational changes in the machinery of government were needed precisely in order to ensure that the UK's indigenous fuel resources 'could be exploited with a more evident sense of urgency'.[109] While the establishment of a new Department may have been intended partly as a sop to the miners,[110] moreover, there was never much doubt about the position which North Sea oil was going to hold in its hierarchy of priorities. Even before the new Under-Secretary of State made his début in the Commons, his Minister had announced on radio and television that 'one of his top priorities will be to speed up North Sea oil',[111] while, as we have noted earlier, this was not an objective from which the soon-to-be-

elected Labour Opposition dissented in any way. Thus the framework
into which the administration of safety was now to be incorporated
was one that, more singularly than ever before, was designed to
encourage the speedy exploration and exploitation of offshore
resources.

With the advent of the Department of Energy and its virtually
complete offshore administrative hegemony for some years thereafter,
this combination of functions within one component of the administrative machine could only exacerbate the risk that the regulatory
system would be contaminated by the increasingly exigent considerations emanating from the political economy of the North Sea. Whereas
in Norway the trend was somewhat different, because of fears that 'a
control function kept within the Ministry [of Oil and Energy] might
have been more exposed to political pressures',[112] in Britain responsibility for offshore safety was located at exactly the point where such
pressures were likely to be most acute. Nor should we forget that the
thrust of such pressures right up through the 1970s was towards the
fulfilment of short-term objectives through rapid exploitation, the
longer-term considerations uneasily embraced by Labour from 1974
onwards notwithstanding. Appropriately enough, it fell to a former
Minister in that Government to voice the objections in principle to
such a structure, albeit after inconclusive, if creditable, efforts to break
the Department of Energy's monopoly had been made:

> What we are saying . . . is that a fundamental principle is being
> breached in an industry that is of great concern to the House because
> the Department of Energy is to be solely responsible for health and
> safety. It has many other responsibilities for the offshore industry –
> to get out as much oil as possible, to get the revenues and, under
> pressure from the Treasury, to consider the interests of the oil
> companies. We categorically and emphatically say that making
> health and safety its sole responsibility is insufficient.[113]

This potential for conflicting priorities within vertically organized
administrative structures is not something that receives much attention in the Noreng analysis referred to above, although its author does
concede that more competition between generalized agencies and the
primary structures would probably benefit safety and working conditions. On other fronts, however, the same author is at pains to stress
that the advantages to be derived from vertical structuring have always
to be offset against fairly evident drawbacks. Thus, for example, he
recognizes that gains in micro-effectiveness may well entail deficits at

the macro level, while the possibility that specialized agenices may develop their own culture and depart from generally established procedures is fully acknowledged. So too is the risk that practices in interpreting and enforcing rules may become institutionally idiosyncratic. Most of all, perhaps, Noreng emphasizes that although vertical structuring does indeed enhance detailed control and access to information, it also means very close contact between controllers and controlled. As a result, it brings in its train the attendant risks of the agency's being 'colonized', of its coming to act as an advocate for those whom it supervises and, indeed, even of control being reversed.[114]

With recognition of possibilities such as these, we are brought back to fairly well charted sociological waters. The tension between micro and macro levels, for example, is a well rehearsed topic,[115] even if the emphasis is rather more upon the possibile contradictions between ad hoc bureaucratic decision making and overall planning.[116] Similarly, the capacity of monopolies to 'occupy' specific parts of the diffuse structure comprising the machinery of the state has often been commented upon, particularly in connection with science and technology.[117] If this is the analogue of Noreng's extreme category of 'colonization', the more general tendency has been cogently described by Habermas:

> It is possible to show that the authorities, with little informational and planning capacity and insufficient co-ordination among themselves, are dependent on the flow of information from their clients. They are thus unable to preserve the distance from them necessary for independent decisions. Individual sectors of the economy can, as it were, privatize parts of the public administration, thus displacing the competition between individual social interests into the state apparatus.[118]

In a subsequent chapter, I shall argue that just such a process of displacement came to play an important part in making the safety of North Sea operations the subject of considerable conflict between different agencies of the British state during the latter part of the 1970s.[119] Before turning to this question, however, it is relevant to map out the way in which some of the above tendencies seem to have permeated the specialized offshore safety regime to a considerable degree in the first place. At one level, I wish to suggest that the special administrative framework which grew up around offshore safety was accompanied by the development of a special regulatory relationship which, while maybe not warranting extreme designation as 'colonization' or 'occupation', certainly did display some highly 'privatized' characteristics. More generally, I will argue that the bond of dependency which has already been outlined at an earlier point also carried over into the realm of safety administration, with important consequences for the nature of offshore safety regulations.

The Development of a Special Regulatory Relationship

For the offshore oil industry, with its own economic commitment to speed, the structure of safety administration described in the preceding section was scarcely inexpedient. As one commentator pointed out in the mid-1970s, with specific reference to the possibility of responsibility for safety ultimately being placed elsewhere than with the Department of Energy, 'There are clear advantages at least to the companies in continuing to deal for the time being with members of the Government Department whose task it is to supervise the rapid development of the resources of the North Sea.'[120] Such 'advantages', however, were neither entirely new nor completely contingent upon some fortuitous coincidence of autonomous interests. Rather, they were the material product of a special relationship which had grown up upon the basis of the unbalanced symbiosis that was discussed in an earlier part of this book.[121]

As we saw in the earlier and more general discussion of the latter topic, one important respect in which the British authorities were over a barrel, so to speak, was that their commitment to speed had generated radical deficiencies in both basic information and expertise. However, just as factors rooted in the political economy of the North Sea exacerbated official difficulties in connection with general information, so too they inevitably had an effect upon the amount and quality of information relating to the more specific issue of safety. Nowhere was this clearer, for example, than in the context of knowledge about the implications of the operational conditions to be expected in such an environment, for here the industry itself was relatively uninformed. Initially, by all accounts, the companies simply extrapolated from their previous experience in other parts of the world, principally from the operations which had been going on for many years in the Gulf of Mexico, and, as a result, they fairly drastically underestimated the magnitude of the problems entailed in operating on the North Sea's Continental Shelf.

While this problem may not have been particularly severe as long as operations were confined to the easier waters of the Southern Basin (which, nonetheless, did see its crop of accidents), the difficulties certainly became acute when exploration moved northwards into the deeper and much stormier waters off the coast of Scotland. There 'conditions which many Americans did not realize existed'[122] led to new problems of design, construction and monitoring as attempts were made to scale up existing technology to meet a more hostile environment. The inadequacy of the basic information available at this stage is evident from the following extract from one report on a symposium held in connection with the proposed Construction and Survey Regulations in 1973:

> There are already insufficient data for the North Sea, and theories are as many in number as there are experts to expound them. Who, then, can be expected to say with certainty which one is correct . . . ?
> Mr Campbell (Shell) reasserted the 'frontier technology' situation of the North Sea and emphasized the lack of data, particularly in predictions of wave height, wave forces, current and therefore stresses and fatigue life of structures.[123]

If the industry itself lacked information of such vital relevance to safety, the authorities were no better off. As Noreng remarks, perhaps generously, the lack of systematic data at this point has to be set against a background of neglect in research and development effort that must be laid mainly at the door of Governments which, 'until the cost escalation reached fairly alarming proportions in 1975–6 . . . seemed to have an uncritical faith in the ability of the oil industry to master the problems of development'. More than that, he argues, in the case of British Governments the balance-of-payments predicament provided good reason for pushing development and for allocating high priority to time targets, with the result (he implies) that there was little incentive to interfere with the companies' practice of foreshortening and even telescoping different stages such as planning, design, construction and installation.[124] The implications for safety data are fairly apparent. As one eminent metallurgist later remarked, albeit in the aftermath of the Alexander Kielland disaster, 'It would be much more sensible if research was done ahead of design, but in the case of oil I believe we are in a great hurry because of the energy crisis and that often the rules are worked out on the sparsest of information.'[125]

If such comments must necessarily smack of wisdom after the event, there were some people who were sceptical enough at the time. According to one critic in 1974, developments like the Brent and Forties fields were a bit like putting Concorde in flight without a flight-test programme, and while he may have been somewhat reassured to learn that Government was currently spending around £1.5 million on research and development, his misgivings about the authorities' role in generating the information requisite to the safety of such undertakings were not entirely allayed:

> It is true that the operators and licensees will have placed upon them, individually and collectively, an obligation continuously to monitor the competence of these structures in operation. There does not appear to me to be any desire on the part of the Government to sponsor research which would seek to anticipate such behaviour. It is essential that we try to get Government-sponsored research and development, in association with the industry, to anticipate the conditions under which these structures are likely to operate.[126]

Such calls for Government–industry co-operation in the acquisition of vital safety information might have been couched in somewhat less optimistic terms if they had taken account of the latter's remarkable penchant for secrecy. To many companies, it must be remembered, one of the great incidental advantages of the North Sea was the opportunity which it offered for the development of an offshore technology that would stand them in good stead when dwindling reserves might increasingly drive them to more difficult maritime locations, and, not surprisingly, they were therefore less than keen to share their growing knowledge with anyone. Thus, for example, it was claimed in 1973 that other firms engaged in 'frontier-technology' developments were pleading with the oil companies to release up-to-the-minute data, but to no avail, while a representative of the Institute of Oceanographic Sciences 'decried oil companies' policy of keeping secret their environmental data'. Significantly, he was also reported to have 'deplored the fact that platforms are being designed using data which is not available for public scrutiny and discussion'.[127]

All this, of course, was some time ago. However, despite the distinct improvements which took place after 1974, there is still reason to believe that an improved flow of information requisite to safety was not all that easily accomplished. Such difficulties were particularly evident, for example, in the case of failures, where information is obviously vital both for purposes of research and for the dissemination of cautionary data. The incentive to maintain production, it has been said, meant that failures might on occasion be quickly rectified but not investigated, with the consequence that the necessary information was not there to be communicated in the first place.[128] As we saw in chapter

2, moreover, such events often went unreported to the Department of Energy, particularly where no injury was involved. Indeed, according to one Certifying Authority, they were not reported 'in a general sense unless they [led] to or [were] part of accidents which cannot escape public detection, e.g. a blow-out or the sinking of a platform'. One result, claimed the same organization, was that institutions engaged in the vital business of fatigue research 'cannot test schemes for crack propagation estimates for lack of examples on which to apply them'. If this is a highly specific instance of informational deficit on the research plane, a more general difficulty became apparent when, in 1978–9, the Burgoyne Committee was unsuccessful in obtaining information about the extent of expenditure by the industry on offshore safety research. As a result, it was left with little more that pious hopes and exhortation:

> The offshore oil industry of the United Kingdom has the opportunity to lead the world in technologies developed to meet the particular challenge of the North Sea. No doubt it foresees benefits from the transfer of such technologies to other countries. We hope the industry will agree with us that this justifies not only a considerable effort of research into the general technologies of exploration and production, but also a proportionate and related research effort devoted to engineering and occupational safety.[129]

Against this background of rapid development, sparse information and industry cards played close to the chest, the authorities had to come very much from behind in terms of marshalling their own expertise with regard to offshore safety. And, inevitably, the process was a slow one. As we have already seen, for example, the first two inspectors were not appointed until 1966 and 1968 respectively, while an identifiable 'inspectorate' did not make its appearance until 1969. By the time that the Mineral Workings (Offshore Installations) Act was passed in 1971, staff involved in safety and operations, including pipelines, still only totalled three. As late as January 1973, the Department of Trade and Industry was still confessing to some worries about the rate at which it had built up the 'technical resources [that]are not required just for the cases we are talking about here, which is mainly assessment of prospects and the evaluation of fields and helping in forming policy, but also for other activities like regulation, safety, inspection and so forth'.[130] By that point, the total staff had risen to seven, but there were still significant gaps, not least in relation to diving, which did not attract its first appointment until the following year. The slow growth of safety expertise within the PED and its forerunners during the 1970s is evident from table 9.

TABLE 9
PROFESSIONAL AND TECHNICAL STAFF
INVOLVED WITH OFFSHORE SAFETY, 1971–9

	Pipelines	Diving	Operations and safety group
1971	1	—	2
1972	1	—	5
1973	1	—	6
1974	1	1	9
1975	1	1	11
1976	1	3	15
1977	1	4	16
1978	3	3	17
1979	5	5	17

Source: Burgoyne Report, p. 89.

Perhaps it was inevitable too that the way in which the authorities gradually made up this lost ground in terms of safety expertise would involve the generation of a particularly close relationship between those whose business was control, including the formulation of regulations, and the industry to be controlled. While the specialized, vertical structure discussed in the previous section may indeed have facilitated a useful concentration of effort in this respect, the inevitable corollary was that the agency was bound to become caught up in a fairly exclusive association with one industry. Furthermore, whatever the latter's significance in terms of economic importance and technological sophistication, it nonetheless remained a highly compact affair, involving relatively few companies and dominated by even fewer. Thus, unlike the horizontal regulatory agencies to be examined later,[131] the PED and its forerunners could scarcely avoid developing a particularly close relationship with one sector of industry.

While the precise nature and implications of this relationship were subsequently to become the subject of bitter dispute, even then protagonists on all sides would seem to be *ad idem* on the appropriateness of a family metaphor. Thus in 1979 the Director of the PED could look back almost with nostalgia to the days of youthful discovery, when 'the industry was in its infancy and both it and the then Inspectorate were learning'. Switching family roles to that of the parent with a wayward child, he went on to describe the current situation as one in which 'the industry is in its lusty youth and may now need the sharp corrective of an occasional prosecution'. As for the effect of those formative years when the rules were still in the making, the diffidence of youth had clearly not been entirely forgotten:

> In the early days they did not have to administer regulations, they had to create them. Because many PED inspectors were involved in the writing of the regulations, there is no doubt that they regard regulations in a different light from the Factory Inspectorate – not as infallible laws which must be obeyed, more as the product of fallible human beings – themselves.

Resort to the metaphor of family was not, however, restricted solely to officials. Shell, for example, invoked a similar image by recalling that 'the Department of Energy Inspectorate has grown up with the offshore industry and has acquired valuable knowledge and experience',[132] while in others such closeness provoked 'fear that this intimate relationship that has developed will inhibit it [the Department] from making criticisms and from imposing costs and other matters'.[133] Nor were such fears assuaged by the fact that there was considerable traffic in personnel between the regulatory agency and the regulated. In one direction, the authorities in many cases acquired their expertise from the most obvious source, inspectors being recruited, as PED later admitted, 'for their experience in either the oil or a related industry'; in the other, and although the Department is obviously coy about publishing the statistics, there was also 'wastage' from the Department of Energy back to the industry, a tendency that was epitomized *in extremis* when the Head of Operations and Safety resigned to become permanent Technical Secretary to the United Kingdom Offshore Operators' Association.[134] From such intercourse would arise charges of an even less wholesome family relationship: 'The Minister should understand that some of us feel deeply that an improper relationship is involved. Indeed, it is an incestuous relationship. The sponsoring Department has no business getting involved with health and safety.'[135]

Although it is obviously impossible to be concrete about the effect which such close relationships, improper or otherwise, may have had on the rule-making process, there is little doubt that they did serve to bolster the industry's confidence in the regulatory stance adopted by the PED and its predecessors. As the general manager of one 'independent' operator explained, the British system was superior to the Norwegian in this respect, since Norway, 'a totally socialist state', had destroyed individual enterprise and had fallen back on looking after its workforce 'from cradle to grave' by means of 'silly little bureaucratic regulations'. In contrast, the British were described as much more sensible because 'the men at the top have experience of the industry and know the problems'. Some way down from the top, PED inspectors themselves seem in the past to have placed considerable value on such experience. According to one of them, there was even a feeling that unless you had worked in the industry you could not purport to know anything about it, much less to undertake research in the area. As for the industry itself, little secret is made of its view that inspectors should ideally be drawn from its ranks. To Shell, an increase in the Department's qualified staff 'with a suitable background of experience in the offshore oil industry' would help to strengthen 'existing links' and to 'facilitate the development of new codes and standards'.[136] Not suprisingly, perhaps, others take a rather different view:

> Unions have pointed out that while it can be argued that the only significant expertise in offshore safety lies with either the Department of Energy or the Companies, the interchange of personnel between the two may raise certain issues. For example, while it is certainly not intended to question the integrity of any individual involved, the possibility of shared values and closed groups amongst offshore personnel may be used as an argument to question the independence of inspectors who have a close association with the industry.[137]

Close relationships, even if described in the more enigmatic terms of 'shared values and closed groups' rather than outright impropriety, did then raise doubts about the extent to which the regulatory agency in question was able to maintain its distance from the industry it had responsibility for controlling. While the speed of North Sea developments posed a constant problem in terms of the adequacy of official information and expertise, the machinery and means whereby the latter were acquired arguably placed offshore safety inside something of a regulatory ghetto. 'Our companies' was the constantly repeated phrase used by one senior Department of Energy official when referring to the offshore oil industry in interview, while Tony Benn later admitted that the tendency for 'the thing to get a bit cosy' had been particularly noticeable during his stay at the Department. More generally, at many points in this research instances were encountered where personnel from the PED did indeed seem to be acting more as the industry's advocates than as its overseers. In one particular pointed case, when members of a commercially organized seminar were told that the offshore safety record was not a good one, it was not the oil men present but the Department of Energy representative who rose to refute this allegation against an industry 'that has come further in seven years than any other comparable one has in three times that time'. Appropriately enough, suggestions that the Department's own approach might be something less than stringent provoked an industry participant to reply that the speaker's remarks were 'less than generous' not only to the industry but also to the Department of Energy.

Close or even closed associations were not the only factor underpinning the growth of a special regulatory relationship during the period with which this chapter is mainly concerned, however. As we saw above, a recurrent theme in debates about offshore safety had been the need to avoid any steps which might hamper the industry's technological progress, upon which depended the continued momentum of North Sea development, particularly as the enterprise moved into more and more difficult terrain. Thus the commitment to speed elevated the industry's technology to almost sacrosanct status, and, not surprisingly, this too was to have a pronounced effect upon the regulatory relationship which became established. In this respect, as in so much else, the authorities found that safety controls were indeed hostage to technological fortune in more senses than one.

At the most obvious level, the way in which the industry was permitted, and even encouraged, to press ahead with operations right up to technological limits where so much was unknown, although not unknowable, meant that safety administration understandably became preoccupied with the prevention of outright catastrophe. As the head of PED explained, ever since the loss of the Sea Gem in 1965, the thinking of relevant Government Departments had been dominated by concern to prevent a major disaster as a result of structural failure, blow-out, fire or explosion, with the possible consequence that 'insufficient attention' was being paid to 'prevention of accidents of an occupational type'. Thus – and while it must be underlined that there is nothing in the British sector's record so far to detract from any claims that might be made for the success of this strategy – the more mundane type of accident, which, as we saw in chapter 3, accounts for the bulk of offshore casualties, was arguably accorded less priority than its incidence merited. Not only that: the 'disaster orientation' of the authorities therefore also came to parallel the industry's own approach, which was always primarily concerned with the more economically and politically consequential possibility of a catastrophe's taking place on its own technological boundaries.[138] In short, the image of offshore danger as an inevitable consequence of technological trail blazing was reinforced by the institutional response to rapid development.

Such preoccupations obviously did little to expose the industry to the healthy scepticism which would insist that on top of the undoubtedly special risks associated with speedy North Sea development, attention should also be focused on a wide range of concerns in which offshore risks are not unique. Nor was the administrative structure already described very likely to have this effect. Perpetuated, if not initiated, by concern to maintain rapid progress, this specialized structure served to confirm the industry's sense of uniqueness by isolating the regulation of offshore safety from developments in other sectors of industry. Thus the input of official experience, in terms of lessons learned from the history of controlling other industrial activities like construction or even manufacturing, seems to have been relatively limited, and the industry was able to continue in the belief that the safety problems about which it had to worry were, above all else, special. Even cross-fertilization from other technologically spohisticated industries seems to have been fairly restricted; the experience of the nuclear, aerospace and chemical industries of risk and safety assessment were particularly under-utilized. More generally, a reluctance to learn from others was noted by one of the Certifying Authorities in 1979:

> Since the Flixborough and other disasters, the interest in safety technology has heightened and more industrial organizations are becoming conscious of the benefits to be gained from its application. Regrettably, the oil industry has not yet generally recognized that the experience gained in other industries in safety engineering can equally well be applied to the offshore oil industry.[139]

But the attribution of a special status to the offshore oil industry, its technology and its unique problems also had another effect, this time upon the rule-making process itself. As we saw earlier in this chapter, anxiety to avoid fettering the industry's spontaneous technological progress had been a major factor behind the decision to leave the specific content of regulations to Ministers, with their power to make Orders in Council. A common enough practice in such legislative areas, its appropriateness had nonetheless been argued with particular force in the case of North Sea oil, and precisely on the grounds of rapid technological change:

> I know that Parliament sometimes likes to feel that it has specified exactly what the regulations should be at the time of the legislation and fights shy of making regulations by Order, but I believe that the Committee will accept that it is right on this occasion that the regulations should be made by Order, because, with this rapidly changing industry, and its advancing techniques and technology, it will be necessary to bring in new regulations and to change them frequently in response to development.[140]

When it came to the making of regulations under the 1971 Act, this deference to the industry's technological status continued to make itself felt. Reflecting the broader problems of imposing detailed requirements in technologically volatile situations, the regulations were mostly written in general terms, with back-up to be provided by Guidance notes. Once again, the rationale was rapid change, while, interestingly, the earlier case for a system that would allow regulations to be changed frequently in response to development seems to have been forgotten. As the PED now put it, 'The fact that the regulations are written in general terms in a broad-brush format obviates *the need* for constant changes to keep pace with technological innovations' (italics added). The system of adding Guidance Notes, it insisted, 'works very well for a fast-moving industry like the international oil industry'.[141]

Much of this is not, of course, to be gainsaid. Similar problems have confronted onshore safety authorities in relation to the use of regulations, and even codes of practice, in connection with requirements likely to be subject to sector variation or technological change, while there is no doubt that Guidance Notes do provide a much more flexible mode of response in a rapidly changing situation. As far as the law relating to the North Sea was concerned, however, one immediate consequence was that the regulatory relationship became one in which attempts to impose detailed legal control over safety matters often became highly contentious. Thus, for example, when regulations *were* on occasion drafted with an eye to the specific, they attracted the disapprobation of the industry, as even in the case of detailed specifications for helicopter accident equipment, medical supplies to be maintained in sick bays and fire-fighting equipment for helidecks. As Esso complained, such detailed requirements were inconsistent with the more general tenor of the requirements in the remainder of the relevant regulations and might therefore be more appropriately put into Guidance Notes.[142] But when the authorities did try in other instances to gain detailed control through such Notes, they found the industry still ready to deploy the same technological arguments and able to remind them that guidance is not quite the same thing as law. Thus Shell later told the Burgoyne Committee of its objections to what was seen as 'an attempt to place the Notes on a quasi-legal footing', to Guidance Notes which enjoined that certain sections '*must* be honoured' and to letters telling operators that they should formally seek relaxation from 'the requirements of the Guidance Notes'. Detailed Notes of this kind, the company claimed, 'can have a restrictive effect of technological development offshore, particularly when they are interpreted as law'.[143] The preferred alternative was clear:

> In our view the principle that must be maintained at all times and in all respects is that Guidance Notes clarify intentions of legislation, but do not bind an operator to any particular course of action. It must be accepted by regulatory bodies that operators may elect to comply with legislative requirements by methods other than given in Guidance Notes. In such instances, it is incumbent upon the operator to justify his chosen compliance route, but equally the regulatory body must give full and fair consideration to it.[144]

In some ways, of course, a statement such as the above is scarcely novel. Those in the know typically resent being told how to run their affairs, while the tradition of denying the relevance and appropriateness of detailed legal controls over industrial practice is a time-honoured one going right back to early factory owners, who were once described as 'men of a warm temperament and a proud spirit who wish to have their own way of doing good, and who kick against any attempt to force them to do good in any other way'.[145] In the present instance, however, the germane point is that within a relationship already characterized by technological diffidence, the detailed application of legal rules had a tendency to become technologically negotiable. Nor is there much doubt that in these circumstances the companies often had the whip hand. In the case of one major North Sea platform, for example, one of the largest multinationals was apparently able to challenge both the drag coefficient specified by the Certifying Authority for the calculation of environmental loads and the 'excessive' criteria laid down by Guidance Notes in relation to wind-induced tides. While no one suggested (or is suggesting) that in this particular instance the result was an unsafe structure, some people see the general context in which such negotiations take place as rather less than satisfactory. According to one chartered engineer who had been involved as a safety engineer in the design of several platforms, the situation left quite a lot to be desired:

> Statutory Instruments [SI: the regulations] and Acts covering the North Sea exploration and exploitation leave too many areas for interpretation. Guidance Notes are not required by law and are only for guidance. The Certifying Authorities interpret the rules differently and require different standards for the same situation. . . . The outcome is that the clients can demand that particular Guidance Notes and Codes of Practice be disregarded because they are not law, resulting in the platform being designed to their requirements of cost and safety. In discussion with the Certifying Authority . . . they agree that only SI shall be used as a guidance. A particular instance recently resulted in the client demanding that the fire protection/fire detection be operated manually only, for one of the largest proposed platforms in the North Sea. Fire codes, Department of Energy Guidance Notes, and all other guidance and codes of practice were not permitted. . . .

If negotiations on this one-to-one level represent one setting in which the industry's technology facilitated the growth of a special regulatory relationship, bargaining over rules and regulations on a more industry-wide basis was another. Thus, for example, in 1974 the Under-Secretary of State for Energy explained that the consultations which had taken place with the industry on the regulations drafted up to that point had been 'as great as with any legislation I have known, mainly because we are dealing with evolving technology and because, in contruction and design, we are working with conditions never before faced when using rigs or building platforms'.[146] Closer to the present, the controversial Guidance Notes on fire-fighting equipment are said to have gone backwards and forwards ten times before finally being accepted, one of the changes being the substitution of 'should be honoured' for 'must be honoured'. According to one company safety officer, they were 'probably the most fully discussed Guidance Notes ever issued', while he was at pains to anticipate obvious criticisms by an appeal to the now familiar canons of technological progress: 'While there may be criticism from some about their lack of definite stated rigid parameters and type of equipment, we have always felt that flexibility is the key note, to allow for technical developments. . . .'[147]

Finally, it must be remembered that whatever the issues at stake, the negotiations surrounding offshore safety laws took place inside a relationship that involved the authorities on one side and, on the other, one of the world's most powerful industries. Nor did the latter comprise just an amorphous array of companies to which officials had to relate on the nebulous basis of projections of what the industry might collectively define as its interest. As far as North Sea operations were concerned, the companies formed themselves into a coherent and effective organization, the UK North Sea Operators' Committee, shortly after the first round of licensing in 1964. The formation of this body was welcomed by Government, which 'increasingly found it a most useful one to consult as providing an appropriate voice for the total offshore oil industry', this notwithstanding the fact that membership was restricted to operators and therby excluded the subcontracting side of the industry, which then, as now, employed the great majority of North Sea workers. This exclusiveness continued after 1973, when the Committee was transformed into the fully registered United Kingdom Offshore Operators' Association, as did a feeling on the part of the British and other 'independents' that the organization was unduly dominated by the large multinationals.[148] Thus negotiations on safety as well as other matters took place within a context where the paramount industry voice was always likely to be that of the large international companies, upon which, as we saw in chapter 4 successive Governments had become so substantially dependent. 'The main channel of communication with the offshore oil industry', reported PED in 1979, 'is through the United Kingdom Offshore Operators' Association . . . who represent all the companies operating in the United Kingdom Continental Shelf.'[149] 'What we do', explained one official, 'is negotiate with the industry: after all, we do still live in a democracy.' Just how such negotiations and the special relationship which they entailed were to be affected by the intrusion of other agents of a democratic society will be examined in the following chapter.

Notes and References

1. Committee of Public Accounts, 'North Sea Oil and Gas', *Parliamentary Papers*, 1972–3, vol. 14, London, HMSO, 1973, p. 23.
2. *Parliamentary Debates* (Commons: Standing Committee A), vol. 1, 1963–4, 19 February 1964, col. 79.
3. T. Daintith and G. D. M. Willoughby (eds.), *United Kingdom Oil and Gas Law*, London, Oyez Publishing, 1977, p. 3.
4. ibid., p. 167ff.
5. *Parliamentary Debates* (Commons: Standing Committee A), vol. 1, 1963–4, 19 February 1964, col. 78.

6. Committee of Public Accounts, 'North Sea Oil and Gas', p. 78.
7. *Parliamentary Debates* (Commons), vol. 688, 1963–4, 28 January 1964, col. 218.
8. *Parliamentary Debates* (Lords), vol. 254, 1963–4, 17 December 1963, col. 238.
9. ibid.
10. *Parliamentary Debates* (Lords), vol. 254, 1963–4, 18 and 19 December 1963, cols. 367 and 391ff.
11. *Parliamentary Debates* (Commons: Standing Committee A), vol. 1, 1963–4, 19 February 1964, col. 78.
12. ibid., cols. 82–3.
13. *Parliamentary Debates* (Commons), vol. 688, 1963–4, 28 January 1964, col. 225.
14. *Parliamentary Debates* (Lords), vol. 253, 1963–4, 3 December 1963, col. 912.
15. *Parliamentary Debates* (Commons), vol. 688, 1963–4, 28 January 1964, col. 220.
16. ibid., col. 224.
17. An exception was made for offshore coal reserves, which were subject to the Coal Industry Nationalization Act of 1946.
18. *Parliamentary Debates* (Commons: Standing Committee A), vol. 1, 1963–4, 9 February 1964, col. 64ff.
19. ibid., col. 80.
20. ibid., col. 76.
21. *Parliamentary Debates* (Commons), vol. 688, 1963–4, 28 January 1964, col. 220.
22. *Parliamentary Debates* (Commons: Standing Committee A), vol. 1, 1963–4, col. 163.
23. *Petroleum (Production) Act 1934*, 24 and 25 Geo. V, c. 36.
24. *Parliamentary Debates* (Commons: Standing Committee A), vol. 1, 1963–4, 4 March 1964, cols. 158–9.
25. ibid., col. 160.
26. ibid., col. 150.
27. *9 and 10 Eliz. II, Ch. 34.*
28. *Parliamentary Debates* (Lords), vol. 253, 1963–4, 3 December 1963, col. 934.
29. ibid., col. 934.
30. See chapter 7 below.
31. *Parliamentary Debates* (Lords), vol. 254, 1963–4, 17 December 1963, col. 230.
32. ibid., 19 December 1963, col. 394.
33. *Parliamentary Debates* (Commons), vol. 688, 1963–4, 28 January 1964, cols. 223 and 277.
34. Clauses 6 and 7.
35. *Parliamentary Debates*, (Commons: Standing Committee A), vol. 1, 1963–4, 26 February 1964, col. 102.

36. *Parliamentary Debates* (Commons), vol. 692, 1963–4, 7 April 1964, col. 885.

37. *Parliamentary Debates* (Commons: Standing Committee A), vol. 1, 1963–4, 4 March 1964, col. 158.

38. ibid., col. 158.

39. ibid., col. 136.

40. *Parliamentary Debates* (Commons), vol. 692, 1963–4, 7 April 1964, cols. 876–7.

41. 1964, c. 29.

42. Daintith and Willoughby (eds.), *United Kingdom Oil and Gas Law*, p. 21.

43. ibid., p. 12.

44. J. Kitchen, *Labour Law and Offshore Oil*, London, Croom Helm, 1977, p. 209, fn. 223.

45. ibid., p. 48; Daintith and Willoughby (eds.), *United Kingdom Oil and Gas Law*.

46. Kitchen, *Labour Law and Offshore Oil*.

47. See chapter 7 below.

48. *Report of the Tribunal appointed to inquire into the Causes of the Accident to the Drilling Rig Sea Gem and other Matters in connection therewith*, Cmnd. 3409, London, HMSO, 1967, p. 1. (henceforward, the Sea Gem Report).

49. ibid., p. 26.

50. ibid., pp. 17–18.

51. ibid., pp. 22 and 26.

52. ibid., p. 18.

53. ibid., p. 2.

54. ibid., p. 24.

55. Kitchen, *Labour Law and Offshore Oil*, p. 135.

56. A. Hamilton, *North Sea Impact*, London, International Institute for Economic Research, 1978, p. 19.

57. Department of Energy, *Development of the Oil and Gas Resources of the United Kingdom* (henceforward, *The Brown Book*), London, HMSO, 1976, p. 39.

58. *Parliamentary Debates* (Commons), vol. 816, 1970–1, 28 April 1971, col. 656.

59. ibid., col. 650.

60. ibid., col. 649.

61. *Parliamentary Debates* (Lords), vol. 315, 1971 18 February 1971, col. 743.

62. *Parliamentary Debates* (Commons), vol. 816, 1970–1, 28 April 1971, col. 648.

63. *Parliamentary Debates* (Commons: Standing Committee G), vol. 4, 1970–1, 17 June 1971, cols. 15–16.

64. See, for example, *Parliamentary Debates* (Lords), vol. 315, 1971, 18 February 1971, col. 742.

65. ibid., col. 746.
66. *Parliamentary Debates* (Commons: Standing Committee G), vol. 4, 1970–1, 17 June 1971, col. 6.
67. ibid., col. 32.
68. ibid., col. 33.
69. 1971, c. 61.
70. Ø. Noreng, *The Oil Industry and Government Strategy in the North Sea*, London, Croom Helm, 1980, p. 102.
71. See chapter 4 above.
72. *Report of the Committee on Health and Safety at Work* (the Robens Report), Cmnd. 5034, London, HMSO, 1972. For a further discussion of this Report, the legislation which followed it and the implications of both for the offshore safety regime, see chapter 6 below.
73. *Parliamentary Debates* (Commons: Standing Committee G), vol. 4, 1970–1, 17 June 1971, col. 5.
74. SI, 1972, 702; SI, 1972, 703.
75. SI, 1972, 1542.
76. SI, 1973, 1842.
77. M. W. Thomas, *The Early Factory Legislation*, Leigh-on-Sea, Thames Bank Publishing, 1948, appendices 1–3.
78. *Parliamentary Debates* (Commons) vol. 867, 1973–4, 16 January 1974, cols. 690–1.
79. ibid., col. 679.
80. ibid., col. 690.
81. ibid., cols. 681–2.
82. SI, 1974, 338.
83. SI, 1974, 289.
84. See, for example, *Offshore Services*, July 1973, p. 17; *Offshore Safety*, (henceforward, the Burgoyne Report), Cmmd. 7841, London, HMSO, 1980, p. 151.
85. June 1975.
86. SI, 1974, 1229.
87. *Parliamentary Debates* (Commons), vol. 891, 1974–5, 30 April 1974, col. 493. These powers were assumed under Sections 26 and 27 of the *Petroleum and Submarine Pipelines Act 1975*, c. 74.
88. SI, 1976, 923; SI, 1977, 835.
89. SI, 1976, 1019; the list is an abbreviated version of the one given in the Burgoyne Report, pp. 84–5.
90. SI, 1976, 1542; SI, 1977, 486; SI, 1978, 611.
91. See chapter 6 below.
92. Noreng, *The Oil Industry and Government Strategy in the North Sea*.
93. ibid., ch. 5.
94. ibid., p. 136.
95. ibid., pp. 141–2.
96. ibid., p. 147ff.
97. ibid., p. 134.

98. Hamilton, *North Sea Impact*, p. 17; for a more swingeing criticism, see the Burgoyne Report, p. 59.
99. The build-up of staff concerned with North Sea safety is set out in the Burgoyne Report, appendix 7.
100. *Parliamentary Debates* (Commons), vol. 788, 1968–9, 13 October 1969, col. 31.
101. *The Reorganisation of Central Government*, Cmnd. 4506, London, HMSO, 1970, pp. 4–5.
102. These and further details of administrative development are, unless otherwise stated, taken from *Civil Service Year Book*, London, HMSO, 1974–9.
103. *The Times*, 9 January 1974.
104. *Financial Times*, 4 February 1975.
105. G. Arnold, *Britain's Oil*, London, Hamish Hamilton, 1978, p. 152.
106. See chapter 6 below. The following information is taken from the *Civil Service Year Book*, London, HMSO, 1979.
107. As a result of the *Health and Safety at Work, etc., Act*, 1974, c. 37.
108. *Parliamentary Debates* (Commons) vol. 867, 1973–4, 9 January 1974, col. 135.
109. *The Times*, 9 January 1974.
110. ibid.
111. *Parliamentary Debates* (Commons), vol. 867, 1973–4, 9 January 1974, col. 136.
112. Noreng, *The Oil Industry and Government Strategy in the North Sea*, p. 141.
113. *Parliamentary Debates* (Commons), vol. 991, 1980–1, 6 November 1980, col. 1493. The speaker was Dr. David Owen.
114. Noreng, *The Oil Industry and Government Strategy in the North Sea*, pp. 113ff.
115. See, for example, J. Scott, *Corporations, Classes and Capitalism*, London, Hutchinson, 1979, p. 156ff.
116. cf. p. 161 above.
117. See, for example, J. Hirsch, 'The State Apparatus and Social Reproduction', in J. Holloway and S. Piccioto (eds.), *State and Capital*, London, Edward Arnold, 1978, p. 101. Interestingly, Hirsch goes on to use the 'oil crisis' of 1973–4 as an example of how the state, despite its function of assuring conditions for the reproduction of capital as a whole, is increasingly forced to 'secure the quite particular profit interests of dominant monopolies'. *loc. cit.*
118. J. Habermas, *Legitimation Crisis*, London, Heinemann, 1976, p. 62.
119. See chapter 6 below.
120. Kitchen, *Labour Law and Offshore Oil*, p. 146.
121. See chapter 4 above.
122. *Parliamentary Debates* (Commons), vol. 867, 1973–4, 16 January 1974, col. 691.
123. *Offshore Services*, July 1973, p. 17.

124. Noreng, *The Oil Industry and Government Strategy in the North Sea*, pp. 95ff.
125. *Oilman*, 5 April 1980.
126. *Parliamentary Debates* (Commons), vol. 867, 1973–4, 16 January 1974, col. 693.
127. *Offshore Services*, July 1973, pp. 17ff.
128. Burgoyne Report, p. 115.
129. ibid., pp. 49–50.
130. Committee of Public Accounts, 'North Sea Oil and Gas', p. 130.
131. See chapter 6 below.
132. Burgoyne Report, p. 137.
133. *Parliamentary Debates* (Commons) vol. 991, 1980–1, 6 November 1980, col. 1493.
134. In its evidence to the Burgoyne Committee, the Trades Union Congress counselled the Inquiry to request information on such transfers of staff. It has not been possible to establish whether this information was ever asked for or received. Burgoyne Report, p. 193. A personal inquiry to the Department of Energy confirmed both that relevant statistics were not available and that such transfers did, in fact, take place.
135. *Parliamentary Debates* (Commons), vol. 991, 1980–1, 6 November 1980, col. 1515.
136. Burgoyne Report, p. 138.
137. ibid., p. 292–3.
138. The Burgoyne Committee, for example, 'formed the impression' that safety research undertaken by the industry did not pay as much attention to everyday occupational matters as to engineering problems. Burgoyne Report, p. 49.
139. ibid., p. 154.
140. *Parliamentary Debates* (Commons), vol. 816, 1970–1, 28 April 1971, col. 648.
141. Burgoyne Report, p. 228.
142. ibid., p. 163.
143. ibid., p. 132ff.
144. ibid., p. 134.
145. N. W. Senior, *Letters on the Factory Act*, London, 1837, p. 32.
146. *Parliamentary Debates* (Commons), vol. 867, 1973–4, 16 January 1974, col. 691.
147. *Oilman*, 23 August 1980.
148. It was this feeling that lay behind the formation of BRINDEX (the Association of British Independent Oil Exploration Companies), Arnold, *Britain's Oil*, p. 78. Similar misgivings led non-British independents to consider forming another splinter group in 1978. *Oilman*, March 25 1978.
149. Burgoyne Report, p. 229.

6

North Sea Oil, Safety and the
Internal Politics of the State

Swindon: *I can't believe it! What will History say?*
Burgoyne: *History, sir, will tell lies as usual.*
George Bernard Shaw,
The Devil's Disciple, Act III

In the previous chapter I traced the development of a special legal and administrative regime which came to dominate the world of offshore safety regulations from the mid-1960s onwards. Accompanied by the growth of a special relationship between controllers and controlled, this regime comprised the main machinery for regulating the safety of North Sea operations during the most crucial phase of the Continental Shelf's exploration and development. To leave the matter there, however, would be to present a highly oversimplified picture, not least because it would project an image of the state as being, at least in this context, somehow condensed into the institutional persona of the Department of Energy and its predecessors. Such a view would be empirically deficient, since it would neglect the fact that even during the peak period of so-called 'vertical structuring' other Departments and agencies were involved in the regulation of offshore safety, albeit in a secondary role. Equally, the picture presented thus far is historically incomplete, inasmuch as the substantial hegemony of the Department of Energy subsequently became exposed to a powerful political and administrative challenge during the latter part of the 1970s. Thus an approach which concentrated more or less exclusively upon the role of that Department would fail to take account of the extent to which the bureaucratic division of labour with regard to offshore safety became the subject of protracted and often bitter dispute between different components of the state itself. Such conflicts, and more specifically the power battle which was to develop between

188   *Oil, Safety and Internal Politics*

the Department of Energy and the Health and Safety Executive (HSE) will form the central focus of discussion in the present chapter.

From a sociological viewpoint, such conflicts between what are often referred to, perhaps infelicitously, as 'state apparatuses' are a matter of considerable importance. For one thing, they index the extent to which the interventionist activities of the state, including its regulatory activities, are riven with internal contradictions stemming from the form of the state itself[1] and from the 'paradoxical' demands which constantly fetter its overall planning capacity and co-ordination.[2] As Miliband remarks in the latter respect, 'The political system allows pressures to be generated and expressed against and in the state, and may turn the state into an arena of conflict, with different parts of the state system at odds with each other, and thereby reducing greatly the coherence which the state requires to fulfil its functions, 'economic ' or otherwise.'[3] On a more general plane, the existence of such conflicts serves as a timely reminder that the regulatory activities of the state cannot be reduced to the operation of a monolith which, by some 'inexorable teleology', secures theoretically pre-given objectives.[4] Representing processes of political struggle within the terrain of the state itself, they underline the fact that, whatever its much disputed determinants, the direction of state activity is neither automatically nor easily established. At the very least, determinants and outcomes are separated by a space occupied by the concrete activities and conflicts in which state apparatuses and their personnel are involved, a space that can be ignored only at the price of adopting the impoverished position which assumes that principles tell us all we really need or want to know.

In analysing the conflict which has arisen between the Department of Energy and the HSE during recent years, I do not intend to adopt such a derivationist approach. Rather, and despite the risk of arraignment for failure to take up a clear position within the extensive and frequently arcane theoretical debate which surrounds the nature of the contemporary state, I wish to come at the matter through concrete institutional analysis.[5] By so doing, I hope that I can both further uncover some of the historically specific constraints which shape North Sea policies, even in the safety context, and show how these are materially realized in the contradictions and conflicts which beset the administration of offshore safety regulations. Equally, I hope to demonstrate that what is at stake here is much more than mere institutional pluralism, a scenario of different bureaucracies promoting their respective interests in the hope of striking it rich administratively where others are doing so economically.

The Politics of Confrontation

The potential for internal warfare over the administration of offshore safety laws has always been fairly substantial as a result of the multiplicity of activities involved in North Sea developments. Inevitably, this meant that a number of different Departments and agencies with already established responsibilities for particular types of operation would become involved, with the concomitant risk of disagreement over demarcation lines, modes of procedure and regulatory philosophy. Of obvious importance here was the Department of Trade's Marine Division, which, on top of a lingering ambition to view everything that goes down to the sea as a ship, had a quite legitimate interest through its responsibilities under the Merchant Shipping Acts.[6] Thus this Department's remit came to include the safety not only of vessels but also of diving operations carried out from British ships and of tow out to permanent location from UK ports, or by British vessels from foreign ports.[7] Some confusion also existed as to its role with regard to British-registered drilling ships when in transit. On one reading of the Mineral Workings (Offshore Installations) Act, such circumstances were covered by section 3(b), which defined an offshore installation as one that 'is maintained, or is intended to be established, for underwater exploitation or exploration'; but such vessels would also normally be registered as ships, and it seems that, in practice, they were treated as falling under the Merchant Shipping Acts when not actually drilling.[8] To confuse the matter further, because the Marine Division had played an important part in drafting regulations pertaining to life-saving appliances and fire-fighting equipment, its surveyors continued to carry out the relevant examinations and certification processes on behalf of the Department of Energy.[9]

While the Department of Trade was for some time the main agency outside Energy to have a function with regard to offshore safety, it was not, however, the only one. Helicopter operations, for example, drew in the Civil Aviation Authority, while on a less obvious level, the use and storage of radioactive materials involved the Department of the Environment and the Scottish Development Department as the bodies responsible for maintaining registers of users.[10] Nor should it be forgotten that with much of the oil lying off the coast of Scotland, there was always a certain tension between laws essentially conceived as 'British' and the fact that Scotland has her own legal system and procedures. Indeed, as we shall see at a later point, the fear that led one MP to 'warn Scotsmen that if they are not careful they will have their jurisdiction ousted on these installations' was not to prove entirely unfounded.[11]

Whatever the potential for conflict embedded in this and other aspects of the overlapping arrangements for the administration of offshore safety, however, it was soon to be overshadowed when the Health and Safety Commission (HSC) (with its executive arm, the HSE) became involved. Comprising the authority which became responsible for most aspects of onshore industrial safety from 1974 onwards, this body was heir to a much longer and different tradition from that developing inside the Department of Energy and, indeed, had been conceived on the basis of a philosophy that was in many ways the very antithesis of the specialized administration which the latter represented. Not only that: its approach could neither as readily accommodate some of the priorities which we saw earlier to have been spawned by the political economy of North Sea oil, nor skirt around the contradictions inherent in the attempt to treat the offshore industry as a special industrial case. As a result, its intrusion into the field of offshore safety became the catalyst for a violent reaction both within the industry itself and inside other parts of the administrative structure. To understand how this came about, however, it is necessary to go back to the circumstances surrounding the establishment of the HSC and the HSE in the first place.

By one of those coincidences which so frequently confound the tidy analytical schemes of historians and sociologists, just as the Mineral Workings (Offshore Installations) Act of 1971 was going through its final legislative stage, a Committee of Inquiry into health and safety at work (the Robens Committee) was reaching the half-way stage in a broader review of the law relating to industrial safety as a whole.[12] Set up in May 1970, this Committee had to contend with a mass of piecemeal legislation that had accumulated on an industry-by-industry basis over the previous 170 years and which, by then, amounted to no less than nine main groups of statutes and nearly 500 subordinate statutory instruments.[13] Given the scale of this task, it is not suprising that the Committee did not report until July 1972, almost exactly a year after the offshore enactment had become law. When it did, its findings and recommendations had far-reaching implications for the way in which the offshore safety regime was developing, even though it had not been able to address that subject in any detail.[14]

Central to the findings of the Robens Committee was the need to create 'a more effectively self-regulating system' within which those causing *and* those working with hazards should recognize their existence and co-operate in taking responsibility for appropriate safety measures.[15] Thus, apart from underlining the duties of management in this respect, the Committee stressed that workpeople should shoulder their full share of the burden and, to this end, 'must be able to participate fully in the making and monitoring of arrangements for safety and health at their place of work'.[16] In all this, the law should play a largely supportive role, to be facilitated by the creation of a unified body of legislation based on what might be called a 'generic hazard' rather than a 'specific industry' approach. Concomitantly, the Committee called for the establishment of a single national authority to be responsible for the law's administration. The fragmentation of such responsibility in the past, it was suggested, had been substantially to blame for the 'obsolescence and inadequacies' of many statutory and enforcement arrangements, since the inevitable overlaps, along with the need for wide inter-departmental consultation, might mean that all could 'move forward only at the pace of the slowest'.[17] As for the predictable claims to special treatment for some industries, the Committee was unconvinced:

> It was suggested to us that safety legislation and administration dealing with mines and quarries, and with agriculture, should remain separate because these industries present certain highly specialized characteristics. It was argued that the mining environment is unique, that the hazards are exceptional, and that in most countries mining safety provisions are separately administered. In the case of farming it was said that farming operations are exposed to natural elements which cannot be controlled, are highly dispersed in very small units, and are characterized by independent working with minimal supervision. . . . We have considered these points carefully. It is, of course, indisputable that many industries have unique features. However, the Factories Act already provides an example of an umbrella statute beneath which special sets of provisions are administered for such diverse and hazardous industries as chemicals, construction, shipbuilding and docks.[18]

In 1974, the recommendations of the Robens Committee were substantially incorporated into the provisions of the Health and Safety at Work Act, which created the unified legislative framework upon which the Report had placed such store.[19] By virtue of Section 2, the Secretary of State was given the power to make regulations which would provide, in prescribed cases, for the appointment of safety representatives from among the employees; these representatives were to be consulted by every employer 'with a view to the making and maintenance of arrangements which will enable him and his employees to co-operate effectively in promoting and developing measures to ensure the health and safety at work of the employees, and in checking the effectiveness of such measures'.[20] No less important, the Act established the HSC and the HSE.[21]

Following the passage of this Act, virtually all of the formerly separate onshore Inspectorates were brought within this ambit of this rationalized administrative framework, though not, it must be said, without a fight. As Mr Michael Foot, Secretary of State in the Department of Employment (to which fell the task of organizing the new system), pointed out at the time, the Robens Report had sparked off a 'first-class Whitehall row' amounting to a 'war which . . . was a classic in Whitehall history'. Although he was happy to disclaim outright victory for his own Department, this gesture of generosity in victory was nonetheless tempered by his unconcealed satisfaction that Employment would henceforth 'retain the sword'.[22] As for the defeated, they could at least derive some satisfaction from having put up stern resistance, as another MP later recalled:

> That Act was not achieved easily. The Department of Energy did not readily surrender its responsibility for the Mines and Quarries Inspectorate. It fought like a tiger, because it did not want to see its empire diminished. The Department of the Environment fought like a tiger against surrendering the Alkali Inspectorate. The Ministry of Agriculture vowed that it would not give up its responsibility for agricultural safety. The battles were fought and, in the main, were won.[23]

In the end, then, victory in the onshore war went largely to the proponents of the integrated approach to safety administration. Protestations notwithstanding, mining and agriculture were both brought under HSC/HSE control, as indeed were nuclear installations – surely the paradigmatic instance of a special case. But what of the North Sea? Although some senior Scottish legal personnel interviewed in the course of this research cling unrepentantly to the view that, by dint of Section 3 of the Continental Shelf Act, the new statute was automatically operative with regard to offshore operations,[24] such was not the opinion officially endorsed in administrative circles. Rather, officialdom appears to have surrendered to the less enterprising, if legally perspicacious, view that the North Sea constituted a residual anomaly to be put right when the occasion offered. In the meantime, primary responsibility remained where it was, with the Department of Energy, and HSC officials could only lament the capriciousness of history:

In a way, I suppose, historically it's unfortunate we didn't have the Robens Committee and the Commission before the development of North Sea oil and gas started, because at the time when the thing started, it was normal practice for specialist Inspectorates to be in the Department which regulated that particular area.[25]

Although Robens had not been able to give detailed consideration to offshore safety, the matter had not however been left completely a hostage to fortune. In a brief but telling paragraph, the prescriptions of which have still by no means been satisfied in recent attempts to justify separate treatment for the offshore industry, the Report had included the Mineral Workings (Offshore Installations) Act of 1971 in a list of enactments which, even on cursory examination, 'prima facie . . . deal with matters which seem to fall naturally within the ambit of the unified system that we propose'.[26] Moreover, the Committee had been quite unequivocal in its view that these statutes and their administration should ultimately be brought into the body of the regulatory kirk, so to speak, 'unless very sound reasons can be adduced for leaving them outside'.[27] Moreover, the ensuing legislation did not entirely rule out such a step. Section 84(3) of the Act stipulated that its main provisions could be extended by Order in Council to cover 'persons, premises, work, articles, substances and other matters . . . outside Great Britain as those provisions apply within Great Britain', while the definition of 'premises' given in the statute specifically included offshore installations.[28] Thus the possibility of placing offshore safety inside the newly constituted machinery was at least on the agenda when the Health and Safety at Work Act of 1974 was passed. What was needed was the political will to implement it.

The first consequential initiative in this respect appears to have been taken by the HSC, which, around the beginning of 1975, made representations that it should assume responsibility for occupational safety on installations.[29] This move sparked off a review of the current distribution of departmental responsibilities with regard to the safety of North Sea operations. At the same time, political pressure was mounting inside Parliament for extension of the Health and Safety at Work Act to the Continental Shelf, partly because of growing concern about the standard of safety offshore and partly, it is said, in the hope of using the provisions of Section 2 to foster unionization on installations.[30] In July 1976, the Prime Minister was asked whether the Government was satisfied with the present arrangements, and, on the basis of the now completed review, he announced that the Act would be extended by Order in Council to cover workers engaged in the offshore oil and gas industry.[31]

Even then, however, the move was somewhat diffident. While the HSC would assume responsibility for the occupational safety 'of workers on all oil and gas offshore installations, including submarine pipelines and pipelaying operations', structural safety and blow-out risks were to remain within the direct remit of the Department of Energy. Moreover, and although the new arrangements for occupational safety were presented as a means of seeing 'that one agency will be responsible for ensuring that common standards of occupational safety are applied both on and offshore', this was not to be accomplished by the obvious strategies of either transferring requisite personnel from the Department of Energy to the Commission's executive branch or simply handing front-line inspection responsibility to the latter body. Instead, an 'agency agreement' was to be worked out whereby the Petroleum Engineering Directorate (PED) of the Department of Energy would inspect offshore installations on the Commission's behalf. The only direct responsibility to be assumed by the HSE appears to have been for some aspects of occupational safety in connection with pipelaying operations and crane ships, and even here the formulators of the new arrangement overlooked the fact that the relevant officials lacked legal powers to requisition any means of getting there! The structural safety of pipelaying vessels was to remain under the care of the Department of Trade, which would also, of course, continue to discharge its duties with regard to ships under the Merchant Shipping Acts. Although the Prime Minister stressed that the new arrangements were designed to produce a 'more rational redistribution of responsibility', the legal and administrative framework for oversight of offshore safety was to remain extremely complex.[32]

This is not the place to rehearse the minutiae of all the difficulties which were to arise in the wake of this important decision, though subsequent relations between the three bodies principally involved do seem to have been characterized by a degree of acrimony unusual even in the competitive world of bureaucratic rivalry. In interview, for example, the then Head of Operations and Safety in the Department of Energy described the Department of Trade as engaging in 'empire building', inspired by the desire to maintain a very large number of marine surveyors despite the marked decline in the size of the British merchant fleet, while the provisions of the Health and Safety at Work Act were merely a 'paper tiger' of no relevance to the situation offshore. Similarly, his successor complained to the Burgoyne Committee that although there was a great deal which the PED might be able to learn from the HSE, the 'big-brother attitude' of the latter and 'their apparently limitless resources of inexperienced personnel' were 'creating poor morale amongst his inspectors'. In the more guarded tone which characterizes HSE pronouncements on such matters, its representatives declined to be drawn into 'discussion of the "machinery of government" aspects of offshore safety', though individual officials did not hide from the Committee their personal view that the HSE 'could and should have been given the complete responsibility for offshore safety'.

Inter-departmental tensions of this kind probably played no small part in delaying the implementation of the new arrangements announced in July 1976. More than a year elapsed before the requisite Order in Council became operative, in September 1977,[33] while the 'agency agreement' referred to above did not take effect until November 1978.[34] Nor, as we shall see at many subsequent points in this chapter, did this step finally put an end to the internecine wrangling within the state bureaucracy. In the meantime, however, events in the North Sea itself had already overtaken the internal politics of safety administration when, in April 1977, a major blow-out occurred on the Ekofisk Bravo platform in the Norwegian sector.[35] Following subsequent discussions between the British Secretary of State for Energy and his Norwegian counterpart, a Committee of Inquiry into offshore safety was set up within the Department of Energy,[36] and the conflicts which had been simmering for some time were set to escalate towards confrontation. Thus even while the state was muddling its fractious way towards a purportedly more rational and better co-ordinated approach to offshore safety, developments were afoot which would render the realization of such objectives doubly problematic.

The Burgoyne Committee was established towards the end of 1978 and reported in March 1980, just some three weeks before the Alexander Kielland, a semi-submersible accommodation rig operating in the Norwegian sector, turned turtle, with the loss of 123 lives.[37] Although the Committee's brief had been 'to consider, so far as they are concerned with safety, the nature, coverage and effectiveness of the Department of Energy's regulations',[38] what was referred to as the current 'tangle of divided responsibilities'[39] was used to justify examination and comment upon the part played by other agencies in this context. Early on in the proceedings, concern was expressed about the establishment of such a wide-ranging investigation, particularly in the light of the recent review and redistribution of departmental responsibilities and the fact that the HSE, in accordance with its customary practice, had set up an Oil Industry Advisory Committee. But the chairman insisted that his Committee's terms of reference 'would obviously have to extend to the areas of interface and the consideration of any overlap or gap', and this position was reflected in a carefully worded memorandum which openly invited evidence on much more than the nature, coverage and effectiveness of the Department of Energy's regulations:

> The terms of reference are designed to cover primarily the Department of Energy's areas of responsibility. . . . Occupational, marine and aviation safety, the present concerns of the Health and Safety Executive, the Department of Trade and the Civil Aviation Authority respectively, are not included, although there are clearly areas of interface and overlap that will have to be considered.

And again:

> What is the present pattern of administration and enforcement? What problems if any arise from the division of regulatory authority between Departments? What objectives does it have or appear to have? Are these the right objectives and do they achieve their purpose? Is the Department of Energy's Inspectorate the right organization for the job?[40]

It would, perhaps, be easy to make too much of the contentious claims that in adopting such an approach the Burgoyne Committee was, quite simply, exceeding its terms of reference.[41] Whether such was the case – and it can be argued both ways – there is, however, no doubt about the consequences of the ambiguity thus created. While the HSC interpreted the original brief as excluding occupational health and safety, and therefore did not make a formal submission (although it did submit a report which hardly justified Burgoyne in asserting baldly that the HSE 'did not submit written evidence'),[42] the broader role which the Committee claimed for itself called forth a barrage of complaints about the existing distribution of regulatory responsibilities. 'The critics of the arrangements described in the Prime Minister's Statement', said the Report, 'comprise virtually all the Operating Companies speaking independently or through the UK Offshore Operators' Association, the two British Certifying Authorities, the Institute of Petroleum, and others', while 'professional bodies criticize the impact of the situation on their particular interests and the

Petroleum Engineering Division of the Department of Energy describes the difficulty of implementing the arrangements.'[43] Moreover, on any reading of the evidence upon which such a remarkable degree of unanimity could be reported, there is little doubt about the strategic target for the attack. As a note of dissent to the main Report observes, 'In spite of the exclusion of . . . the role of the HSE from the terms of reference, virtually all the organizations associated with the UK Offshore Operators' Association chose to aim their attacks on the HSE.'[44]

In the light of such overwhelming criticism, the Burgoyne Committee recommended that, as a matter of urgency, responsibility for all offshore safety matters should be placed in the hands of a single agency.[45] Borrowing an appropriately pugilistic metaphor, the Report concluded that the Department of Trade was 'not a contender' in this respect, and that the remaining choice, 'entirely a matter for the Government', must lie between the Department of Energy and the HSE.[46] The respective advantages and disadvantages of each course of action were then subjected to some fairly cursory discussion – the whole import of Robens, for example, seems to have been effectively ignored – but in the end, the diffidence of indecision gave way to a forthright recommendation: 'We conclude that there is a strong case for the single agency to be the Department of Energy and that the Department of Energy would be capable of performing the task.[47]

With the publication of this Report, the battle for control over the administration of occupational safety regulations in the North Sea seemed to be approaching a near rout. Quite clearly, implementation of its recommendations would bring to an end the responsibilities for offshore safety which the HSE had only recently assumed and which, indeed, had hardly been put into practice before being challenged. Interestingly, for example, when the UK Offshore Operators' Association was challenged verbally to explain why its complaints had not been brought to the HSE's own Oil Industry Advisory Committee, on which it is represented, the reply was the rather lame one that this was 'a very new organization that had only met two or three times and had had little time so far to get into details'. Against the background of such precipitate disenchantment, it is difficult to avoid the impression that, by 1980, even the limited and short-lived intrusion of the HSE into the affairs of the North Sea had touched some very raw nerves indeed. Exactly which nerves these may have been is the central question to be addressed in the remainder of this chapter.

Disharmony through Harmonization

When the Robens Committee of 1972 came to consider how its recommendations might be affected by British membership of the EEC, it noted that Articles 100–102 of the Treaty of Rome 'provide for the removal of technical obstacles to trade through the harmonization, by means of directives, of regulations, standards and administrative practice'.[48] Although only two such Directives with a bearing on safety and health at work had been adopted at that point, the Committee observed that several others were in an advanced stage of preparation and that further initiatives might be expected under Articles 117 and 118, which contain provisions for the promotion of improvements in working conditions. Far from being out of step with such developments, moreover, Roben's own proposals were seen as facilitating 'the process of harmonizing national standards with EEC standards where necessary'. A broad enabling statute of the kind suggested would permit modification of detailed standards without extensive legislation being required, while the case for a single national authority was strengthened by EEC membership:

> it is possible that in the longer term the European Commission will increasingly become a centre of initiative for the promotion of safer working conditions and practices generally. As a member of the Community Britain will, of course, play a full part in the formulation of any relevant provisions. It seems reasonable to think this will be done more effectively if our own legislative and administrative arrangements for safety and health at work are, as we have recommended, co-ordinated under one national Authority. In our view, therefore, entry into the European Economic Community reinforces the case for a single central Authority in this country responsible for matters concerning safety and health at work.[49]

In the wake of Robens and the Health and Safety at Work Act which followed, HSC/HSE involvement in the process of international harmonization became quite substantial. By 1976, the HSC's chairman was able to report that his organization had been providing the 'link between Government and industry in developing positive United Kingdom health and safety policies and the response to proposals generated by other Member States' under the auspices of Article 100. Equally, he could commend his Executive for its involvement in the EEC's Advisory Committee on Health and Safety at Work, and could conclude, with perhaps just a touch of British disdain for the ways of foreigners, that in both these contexts 'Our aim is to ensure that the standards adopted throughout the Community are no less stringent than those we intend to apply within Great Britain.'[50] In its turn, the Executive explained that it had continued to play an important role in international discussions (a claim certainly justified by the fact that, by that point, it was the 'lead' British agency in connection with no less than seven adopted and nearly fifty draft Community Directives),[51] and that 'we are obliged to commit resources to the discussions in order to protect UK interests and to ensure that the Executive does not develop policies seriously at variance with the consensus of opinion in Europe.'[52]

Against this background, the extension of the Health and Safety at Work Act to the Continental Shelf in 1977 was fraught with serious implications for the coherence of the British state's approach to the control of offshore safety. From the outset, Britain has jealously guarded her jurisdiction in this context, and indeed, has arguably exceeded her entitlement under the Geneva Convention on the Continental Shelf (1958) in several respects,[53] a measure of self-indulgence which may have a considerable bearing upon the reportedly strenuous efforts which were made in 1979 to prevent an injured worker, Stephen Filby, from pursuing an industrial injury benefit claim before the European Court of Justice.[54] Britain's position is, in effect, that North Sea oil is none of the Community's business. For its part, however, the European Commission takes the view that 'the Continental Shelf is to be treated, with regard to the applicability of the Treaty, like the territories of the signatory states', and, accordingly, it regards its Directives as being 'applicable to the relevant national measures and practices as they relate to the Continental Shelf'.[55]

Here, then, there was plenty of room for conflict. As we shall see below, the HSC and the HSE were committed to harmonization of regulations on an industry-wide-basis – through a 'generic hazard' as opposed to an 'industry specific approach' as already mentioned – and were therefore keen to incorporate offshore safety, where appropriate, into the more general regulatory framework applying onshore. Since, however, the latter entails some degree of harmonization with EEC Directives, the North Sea could be drawn into the process by proxy. Thus, while the jurisdictional issue might not be conceded in principle, creeping harmonization could compromise it in practice. Problems arising from the use of 'standards' in EEC Directives, for example, might indeed be technically confined to the onshore scene, at least on the British Government's idiosyncratic reading of the jurisdictional issue, but unless the HSC and the HSE surrendered to the industry's demands for separate treatment, it is difficult to see how this convenient separation could be operationally sustained for very long. Such dangers epitomize not only the problems confronted by the nation state in its attempts to maintain a relatively autonomous position within the wider spectrum of an international economic system, but also the difficulties which it encounters in this respect, as in others, from the pronounced lack of co-ordination between its own component bureaucracies.

In the short term, however, it was not the jurisdictional implications of HSC/HSE commitment to harmonization on an international level so much as the fact that these exacerbated other problems arising from generic harmonization on the domestic plane that touched the more sensitive spot within the regulatory relationship. (Indeed, it must be conceded that although the Burgoyne Report did mention 'a trend' towards international harmonization, it chose the UN Intergovernmental Maritime Consulative Organization rather than the EEC as exemplifying the tendency).[56] This convergence between the two issues was noted by Dr. Burgoyne himself at one point in his Committee's proceedings:

> It would appear logical that health and safety offshore should ultimately become the responsibility of the Health and Safety Commission. This would presumably mean that the relevant legislation would be the Health and Safety at Work Act and regulations made thereunder; and that the responsibility for enforcement would be that of the Health and Safety Executive. . . .
> The transition to this position would be a lengthy one. The development of regulations under the Health and Safety at Work Act, especially if HSE insists (as they appear to do) on comprehensive harmonization with UK industry generally and with the EEC, would take many years. Can the needs of the offshore industry possibly be served in this way during, say, the next five years?[57]

In this statement there are several points of significance. One, to which we shall have cause to return later, is the way in which the issue is portrayed in terms of a short timescale; another is the interesting, though perhaps inadvertent, representation of the problem as having to do with the industry's 'needs'. But the central point is that the chairman here accurately identifies what has been the HSC/HSE approach to the regulation of industry in general since their inception. In its first Report, for example, the HSC had stated very firmly that although much of the Executive's programme concerned specific proposals to deal with particular hazards, 'we do not intend to make piecemeal amendments to a body of legislation which has grown up throughout the last century to deal with particular situations in specific industries.'[58] Nor, as the following extract from an interview with the chairman and another senior HSC official shows, was the North Sea oil industry going to be treated automatically as an exception to the general rule:

> Well, I think it's true to say, isn't it, that going through a phase of looking at all the existing legislation which applies onshore, progressively – it can't be done overnight, of course – but we are progressively updating it, where you have got one code of regulations, say for mines and quarries, and another one for factories, which cover the same ground, trying to bring them together into a single across-the-board one. Then we will ask the question each time one of these comes up, 'Well, should this be applied offshore?' 'Would it be appropriate for the hazards, like electricity . . . or dangerous substances?' On the face of it, it seems very appropriate . . . that we should apply the same standards as you use onshore. But I think, by reason of the workload that is involved for the Commission and the Executive, it's going to be done as part of the review which is already in operation. It will only be in the specific areas, like diving, where obviously there are special offshore risks that we will in the early days want to take specific action outwith the general programme. . . .

Such an integrated approach to offshore regulations was likely to annoy both the industry and the Department of Energy, since it jarred with the central assumptions underpinning their relationship – that the North Sea was a special case with special needs which could not be adequately catered for within a more broadly based regulatory programme. Thus, on the industry side Esso warned Burgoyne that one consequence of 'overlapping and duplicative regulations [an oblique reference to HSC and HSE involvement] is the substantial effort required merely to sort out and find ways to comply with the various regulations, seek exceptions and attempt to have inappropriate regulations changed'. The interests of safety would be better served, the company asserted, 'if these efforts could be directed towards enhancing safety under clear and sound regulations, administered by a single regulatory agency'.[59] Similarly, the UK Offshore Operators' Association offered examples of instances in which the involvement of the HSC/HSE and the Oil Industry Advisory Committee had caused delay both in the formulation of regulations and, no less important, in the promised modification of some about which the industry had expressed concern. On the first score, for example, the HSE had asked for comments on proposed legislation covering pressurized vessels by the end of May 1978, but was now indicating that a second consultative document would appear at the end of 1979, a delay which confirmed the Association in its view that 'the process of consulting with all parties onshore . . . is obviously a major undertaking and time-consuming.' As for promised modifications, in 1976 the UK Offshore Operators' Association had been assured by the Department of Energy that the new Operational Safety, Health and Welfare Regulations (SI 1019) would be altered; but, following the extension of the Health and Safety at Work Act, the HSE and the Oil Industry Advisory Committee had become involved, little progress being made thereafter. Once again, the point was made that 'this delay would not have occurred if the re-drafting had been left with the Department of Energy.'[60]

If the industry was annoyed by the more broadly based approach of the HSC and the HSE and their reluctance automatically to accord unique status to the offshore situation, the PED of the Department of Energy was scarcely any less incensed. In a memorandum to the Burgoyne Committee, for example, the Directorate's head conceded that there might be much merit in having, as far as practical, the same standards of occupational safety both on- and offshore. 'However,' he insisted, 'the industry is different from any other in the UK because of the multiplicity of activities carried out on a single installation, apart from the special problems associated with the hydrocarbon reservoir.'[61] Similarly, the PED carried the battle over the Operational Safety, Health and Welfare Regulations into the enemy's own camp by proposing to the Oil Industry Advisory Committee that the issue should be dealt with by enlarging the existing SI 1019 standards 'as a means of giving guidance to employers and workpeople on more effective methods of accident prevention'. In the power struggle which ensued, this approach was 'obviously favoured' by the employers as against the HSC's intention to start a detailed programme of Guidance Notes which would be issued under the general requirements of the Health and Safety at Work Act. Most significant of all, perhaps, in its formal evidence to the Burgoyne Committee, the Directorate explicitly en-dorsed the orthodox view of the industry as being 'by its very nature a relatively short-term one' and, against this background, went on to echo the operators' misgivings about the role of the HSC:

> The pattern of responsibilities for offshore safety is complex. While it is appreciated that one Government body cannot and should not

act in isolation from another interested organization, the very broad scope of the HSC's activities and their policy of attempting to standardize regulations, codes of practice, etc. across all work activities means that any proposals they make take a long time to come to fruition. Long timescales are particularly wasteful in what we have already said is a relatively short-lived industry.[62]

With a statement such as this, we come close to the nub of the harmonization argument. What was at issue was much more than an inordinate degree of bureaucratic dilatoriness born of a cumbersome administrative machine with wide responsibilities and unwieldy procedures; it was the question of whether the framework and procedures adopted with regard to onshore industry were compatible with a particular definition of the offshore enterprise. From the industry's point of view, there was little doubt that they were not. Anxious to be fettered by nothing more than its own technological limits, the offshore oil industry sees statutory safety controls in general, and harmonization in particular, as impeding its spontaneous progress. The point was cogently expressed to the Burgoyne Committee by BNOC:

> Legislation should be kept to a minimum and should be expressed in simple and general terms. The majority of the control should be obtained by the use of regulations, codes of practice and guidance documents. . . . These documents must be written specifically for offshore operations, although standards and procedures used in other fields may be drawn upon in the drafting. Attempts to co-ordinate the contents of these documents too closely to practices in other fields should be resisted, and the temptation to introduce their provisions into statutory legislation should also be resisted. This is important because in a continuously developing industry, flexibility and rapid action is required if an adequate response to experience and advances in technology is to be achieved.[63]

As we have already seen, this view of the North Sea oil industry as one within which the pace of technological and other developments should not be unduly constrained by law had been a recurrent theme in earlier legislative debates on offshore safety.[64] As long as the Department of Energy carried the bulk of responsibility for the formulation and implementation of regulations, the maintenance of a 'desirable' balance in this respect had been relatively unproblematic, since the industry's self-definition was institutionalized within the bureaucratic structure. When the HSC and the HSE came on to the scene, however, this definitional consonance could no longer be taken for granted quite so readily. Geared to harmonization, these bodies were more sceptical about the uniqueness of many aspects of offshore technology[65] and, accordingly, were less inclined to accept the case for special treatment without question. The result was that their approach to the regulation of offshore safety was indeed slower, less flexible and, some would say, more considered. Thus by 1980 an institutional rift had developed within the bureaucracy, a rift which had potentially far-reaching implications for Britain's relationship with the North Sea oil industry. As far as harmonization itself was concerned, the Burgoyne Committee may have the closing word – though, it is to be hoped, not the final one:

> The offshore industry is not routine; it includes some routine features but these are put together in a variety of ways to meet the need for rapid and innovative development of installations for different locations. Flexibility of approach, speed of reaction and individual treatment of each case are therefore required in dealing with the problems encountered.
>
> Speed of response and flexibility of approach are more likely from an organization with only one industry whose safety matters are its concern. Conversely an organization with responsibility for the majority of industrial safety matters will naturally tend to view any individual industry in a wide context and formulate guidelines and regulations applicable across the whole of industry. The consequences of such an approach are greater rigidity and a slower response.[66]

De-privatization and Independence

From what has already been said in this chapter, it will be clear that the involvement of the HSC and the HSE in the regulation of offshore safety posed a substantial threat to the special relationship which had previously existed between the industry and the Department of Energy. At its simplest, the intrusion of these organizations meant that companies now had to deal with additional regulatory bodies and that, perhaps understandably, they came to resent the inconvenience which this involved. As the PED succinctly put it, 'Operators had become accustomed to dealing in the main with one Government authority in their offshore operations, the Department of Energy, and they have not welcomed another.'[67] Nor was it just the proliferation of control agencies that occasioned annoyance at this level. Whereas the industry had formerly been able to count upon the fairly undivided attention of the Department of Energy, it now had to contend with organizations carrying responsibilities for a much broader spectrum of industrial activity. In consequence, an element of competition for attention was introduced into the situation and further exacerbated the industry's sense of frustration. As Shell somewhat petulantly complained in connection with intercommunication on safety technology, 'The HSE is more remote from offshore activity and has to deal with many more competing claims from other sections of industry.'[68]

Another respect in which the PED was certainly much less 'remote' from its clientele concerns the recruitment and deployment of its staff. As we have seen in a previous chapter, personnel were recruited in the main for their experience in the oil or a related industry, and, inevitably, given the PED's size and its specialized brief, there was little scope for mobility.[69] In contrast, the Factory Inspectorate – the HSE component with the most immediate involvement in offshore safety – moves most of its inspectors around the country every five years or so, in order to avoid undue identification with local industry, and operates a much broader recruitment and training programme which is not nearly as dependent upon the industries which it regulates. In the latter context, the gap between the two bodies was made quite apparent by one HSE man:

> It's not magic; we train all sorts of people to do this. We've got a guy being transferred up here who is a historian by background, and, you know – I almost said you can . . . train anybody, but that's grossly unfair – but, you know, we get in this job quite a number of people who are sort of 'graduates' . . . and they make confident inspectors without any bother. In fact, after the initial stages you wouldn't be able to tell whether they were arts graduates or graduates in heavy engineering industry. . . .

Such a statement raises, both directly and indirectly, some crucial differences in the HSE approach, differences which were to prove fairly unpalatable in the offshore context. To the HSE, for example, transfer of personnel may indeed be a strength, but to the North Sea operators, it betokened quite the reverse. According to the Technical Secretary of the UK Offshore Operators' Association, himself formerly Head of Operations and Safety in the PED, such mobility was not a virtue but a source of 'continuity problems arising from the constant nmovement of staff'. Similarly, there is no suggestion here that the capacity to carry out every major inspection function with regard to a particular industry depends upon specialized expertise or previous experience within it, whereas the offshore industry is said to have little respect for 'non-practitioners'. The implication was stated pretty bluntly by BP when the company told Burgoyne that 'what few experts there are outside the Industry and its related professional bodies are probably located in the Department of Energy.'[70] Moreover, it is crucial to understand that the kind of inspector described above plays an important part in an operational structure which, in addition to acknowledging the need for specialist teams which can be called upon when appropriate, emphasizes the role of the general inspector with a broad brief for occupational safety. Thus, the issue became the extent to which the offshore oil industry, far from being a secret world comprehensible only to the insider, might in fact be better regulated by general inspection with specialist back-up. Significantly, just such a suggestion that occupational safety might be better catered for by 'Factory Inspectorate type of inspectors, with the PED inspectors providing specialist advice' was greeted less than enthusiastically by the head of the latter body:

> PED believes that this would be a retrograde step and, as it would in no way remove the necessity for the specialist inspectors to visit the offshore installations to carry out this part of their function, it would lead to duplication of effort on the one hand, and confusion and objections from the oil companies on the other. Transport to the offshore and accommodation there is a very real major problem, particularly during the early days of construction and drilling when every seat and every bed is at a premium. Also the oil companies would tend to be at a loss and would have difficulty in appreciating the fine distinction between the two classes of inspectors.[71]

Duplication of regulatory agencies, different patterns of staff recruitment and deployment and a potential for the 'de-specialization' of large sectors of offshore safety all bade fair to impinge adversely upon the character of the Department of Energy's relationship with the oil industry in the years following 1977. Important as such irritants may have been in promoting the backlash against HSC/HSE involvement in the North Sea, however, there was also something else at stake, namely, the extent to which the administration of offshore safety would be allowed to become independent of the requirements of production. More specifically, the advent of these bodies threatened the Department of Energy's special relationship at its most vulnerable, if crucial, point – the fact that it encompassed not only the issue of

safety, but also other vital matters such as licensing, the encouragement of exploration and the maintenance of progress towards self-
sufficiency in oil. Embodying a movement, however tentative, towards
the separation of these other functions from the administration of
safety, the involvement of the HSC and the HSE in the regulation of
offshore safety threatened to imbue the latter with a degree of relative
independence which it had formerly lacked.

When the Robens Report suggested that statutes such as the Mineral
Workings (Offshore Installations) Act should ultimately be brought
within the unified framework being proposed 'unless very sound
reasons can be adduced for leaving them outside',[72] one such reason
for exclusion had already been considered and rejected by the Report's
own argument in another connection. It had been suggested to the
Committee that one good reason for leaving mining and agriculture
outside the new arrangements was that these industries might be best
regulated within the context of other government policies administered by their 'sponsoring' Departments.[73] But the Committee
has been unimpressed:

> We see little of real substance in this argument. In the first place, it is
> not generally applied. If it were, the Factory Inspectorate might
> appropriately be divided amongst the various divisions of the
> Department of Trade and Industry and the Department of the Environ-
> ment, which are 'sponsoring' Departments for most of the industries
> covered by the Factories Act. In the second place we saw little
> evidence that the activities of the full-time safety inspectors for
> mining and for agriculture are closely meshed in with the general
> work of their parent Departments. . . .[74]

Despite the force of Robens's reasoning in this connection, when the
Health and Safety at Work Act was finally extended to the Continental
Shelf in 1977, it was the sponsorship argument that was deployed
successfully to justify at least partial exclusion of the North Sea from
the unified framework. Asked why complete responsibility for
offshore safety had not been passed to HSC/HSE at this point, the
latter significantly explained that 'a special relationship was perceived
between Energy as the Sponsor Department and the industry.' As a
result, it was deemed appropriate 'to use and maintain within Energy,
the engineering expertise to cover oil-well and structural engineering
safety'. Nor, it was admitted, was such deference to special relationships entirely unique. When the Mines and Quarries Inspectorate had
been transferred to HSE, a somewhat similar situation had been

acknowledged, and in consequence, 'Energy as well as Employment needs to be consulted before new safety regulations are made.' It is worth noting, however, that this requirement notwithstanding, the relevant Inspectorate, unlike the PED, *had* actually been transferred.

One answer to any Robens-type objections to the above arrangement, or to the agency agreement whereby the Department of Energy retained front-line inspection responsibility even where HSC/HSE were in charge of policy, is that the safety personnel within the PED are indeed 'meshed in' quite closely with the other oil-related work of the Department. On an administrative plane, this is fairly evident from the fact that the chain of command in which they are located runs up through the Division's director to the same Deputy Secretary who is responsible for the licensing and general North Sea policy work of the Petroleum Production Division.[75] Nor is there any dearth of observers who suspect that such broader considerations are not easily ignored. According to an officer of one trade union, for example, 'without a shadow of doubt' there are divided loyalties and conflicts as a result of involvement in the 'economic and production side'; while another described the enforcement process as taking place in the knowledge that 'over their shoulder there is a Government which is determined to get the oil out of the North Sea as quickly as possible'. HSE personnel were equally uneasy:

> the danger is always there, I think, that, you know, the production side of it might, well, you know. . . . There's good reasons why they've got to produce this oil, you know – the good of the country and all this stuff. And that's pressure on the inspectors as it is on anybody else. . . .

Even Burgoyne saw the danger clearly enough, even if it was ultimately discounted:

> The removal from the Department of Energy of any responsibility for safety matters would satisfy the argument that as a matter of principle no single organization should be responsible for safety matters and sponsorship of the industry. The argument goes that it is inevitable that economic and political pressures will influence attitudes towards safety, and in individual cases undue risks may be condoned for the sake of economic benefit.[76]

Not surprisingly, any suggestion that the Department of Energy might be inordinately less than independent in its approach to offshore safety is roundly dismissed in the relevant official circles. Faced with such an inference, for example, the head of the Directorate insisted that 'In no cases as far as the Department of Energy is concerned has production been allowed to override safety considerations, and if the situation should arise where production was so vital that safety had to be jeopardized, then a ministerial decision would be needed.' Appropriately enough, the junior Minister responsible for offshore safety within the Department was equally adamant. As he explained in a television interview:

> I don't think that this necessarily means that because we are anxious to reach self-sufficiency, which we shall do later this year, that in any way we are going to sacrifice safety standards. Everybody in the Department of Energy, from the Secretary of State downwards, is dedicated to safety as a number one priority. We realize the value of the resources we have in the North Sea; we are anxious to have those reserves brought to the surface as quickly as possible; but we are not going to sacrifice one life in order to achieve this, if we can possibly help it.[77]

In a subsequent chapter we shall see that such bland assurances do not accurately reflect the relative priorities which have, at least on occasion, been alloted to production as against safety in the enforcement of offshore safety regulations by the Department of Energy.[78] Here, however, the point to be made is that, despite the understandable protestations, PED personnel do seem to be caught in a much tighter double-bind than their HSE counterparts. In short, given their Department's responsibilities for the implementation of oil policies in general, the argument that they cannot easily distance themselves from the broader economic and political implications of the North Sea remains substantially unanswered. On the contrary, the possibility that the HSC and HSE might achieve this in somewhat greater measure seems to have constituted one of the main objections to the extension, or even the continuation, of their offshore role. In the former context, when a PED man was asked why complete responsibility should not now be assumed by the HSE, his telling reply was that 'They could not balance the exigencies of safety against the requirements of production as we can.' *Mutatis mutandis*, the need for broader considerations to be taken into account also figured prominently in the case for reversing the arrangements and restoring the Department of Energy's offshore suzerainty. When the Burgoyne Report posed the rhetorical question 'Why should the Department of Energy be given back entire responsibility for safety matters which it had until 1977?', it was able to cite the

environment, the remoteness of operations and the concentrated location of diverse activities as distinguishing features of the offshore
industry.[79] But it very quickly fell back on one central argument to
justify separate treatment:

> Accepting that the offshore industry is quite different from any
> encountered in normal UK industrial life, we now turn to the reasons
> it is claimed to need separate treatment. The UK economy derives
> great benefits from the resources being exploited on the UK Shelf.
> There are differences of opinion as to how these benefits can be
> maximized but virtually all commentators start from the common
> assumption that the oil in particular is a central factor in the UK's
> economic performance. The offshore industry is not routine. . . .[80]

The 'very sound reason' being adduced here was, then, an economic
one. Moreover, far from canvassing independence from economic
considerations on the part of the enforcement agency in question, it
was a justification which asserted by implication that the economic
centrality of oil could be accorded greater prominence within a separately administered regime for safety. To be sure, with growing production and concomitant government concern about conservation,
the logic of the argument did not necessarily mean an unqualified
return to the good old days. But it did entail reversion to a situation in
which economic imperatives, however defined, could be more readily
accommodated within a separate and specialized relationship. Conversely, the fact that the argument should both be couched in these
terms and coupled explicitly with the case for the Department of
Energy suggests that the HSC and the HSE were not seen as offering
the same guarantees in this context.

At this point it should be stressed that I am not here claiming
complete autonomy from overarching economic considerations for
the system of safety administration which regulates the vast majority
of onshore industries. The history of the relevant legislation over the
past century and a half would certainly not sustain such an uncompromising claim,[81] any more than recent reports of pressure on the
HSE to be more flexible in its general approach 'because of the
economic climate' would accord with such an assertion in the contemporary context.[82] Equally, it would be unrealistic to suppose that
HSE personnel concerned with offshore matters could totally escape the
kinds of pressure under which their PED colleagues have to operate;
nor, indeed, do they deny it. Even the most ardent advocates of

complete HSC/HSE control over offshore safety endorse the view that 'it is clearly impossible for any regulatory body to pursue its objectives in total autonomy from either the interests of the industry it regulates, or the perceived requirements of the economy as a whole. . . .'[83]

Having conceded this much, however, it must also be said that the matter is one of degree, and that there are good reasons for attributing a greater measure of relative independence to the Commision and its Executive. At the most obvious level, the fact that they are institutionally separated from Departments with primary responsibility for the management of the economy's different sectors means that they are not so immediately prone to the pressures generated by broader departmental purposes, even if they can never be entirely immune. As one HSE inspector remarked in interview, 'A divorced health and safety function . . . makes life for the inspectors far easier.' More important, as suggested in the first chapter of this book, the ideological significance of factory inspection in historical and sociological terms has lain precisely in its capacity to maintain such a stance of relative independence, thereby demonstrating that the state does not leave the physical conditions under which labour is performed to be merely the unalloyed consequence of market forces, employer's whim or overriding economic imperatives.[84]

To this uneasy legacy the HSC and the HSE are substantially the heirs, and, not suprisingly, it shows up in their attitude to offshore safety. Interviewed at the time when extension of the Health and Safety at Work Act to the North Sea was imminent, the Commission's chairman pointed out that they had already been through an analogous argument with the local authorities, who had formerly 'enforced the law in the offices where they, themselves, are employers'. Glimpsing something of the ideological issue, he described such a situation as one in which 'justice is not being seen to be done', and went on to draw a parallel with the North Sea. Ideally, he suggested, a Department with such direct involvement in production (the Department of Energy's links with the operations of BNOC were the case in point) should not make safety regulations as well; even if it did, the inspectors would still gain 'a little bit of independence if they were all attached to a central unit like ours, whose job is inspection and enforcement action'. As for the broader economic considerations, his answer was quite unequivocal: 'I don't see why our policy needs to be influenced by either the price of oil or even by the economic considerations involved in it. . . .'

What I am suggesting, then, is that even the limited intrusion of the HSC and HSE into the world of offshore safety presaged a new degree of independence in the formulation and enforcement of regulations, and that this in turn, played no small part in generating the backlash so apparent in the Burgoyne Report. Relative as it may be, such an approach would not ensure adequate cognizance of the unique posi- tion which North Sea oil, rightly or wrongly, had come to occupy in the management of the crisis-ridden British economy. Thus, on top of the unacceptable implications of harmonization (themselves not economically inconsequential) and the potential de-privatization of the Department of Energy's close relationship with the industry, the basic economic nexus underpinning North Sea development imposed additional constraints upon the extent to which the HSC and the HSE would be allowed to assume an active role in the administration of offshore safety. That such should have been the case underlines the fact that we are here dealing with something much more than bureaucratic rivalry and institutional pluralism. As Miliband remarks, 'Whatever the state does by way of provision of services and economic intervention has to run the gauntlet of the economic imperatives dictated by the requirements of the system – and what emerges as a result is always very battered.'[85] In the case of the North Sea, since oil has come to be seen as an at least temporary lifeline providing a breathing space for the restructuring of the British economy, economic imperatives have been accorded an even higher than usual priority within the politics of the state. What is likely to be 'battered' as a consequence is a structure of safety administration which is not as immediately responsive to such demands.

Collectivism and Participation

Just as the involvement of the HSC and the HSE in regulating offshore safety potentially cut across the economically unique status being accorded to North Sea oil in the management of the crisis confronting the British state, so too was it out of line with another fairly unique feature of the offshore industry – the relative weakness of the organized labour movement. In conveniently symmetrical fashion, while the oil is perceived as something of a panacea for the nation's economic ills, the industry which exploits it is distinguished by its low rating on some crucial indices of what has been called 'working-class defensive capacity', arguably one of the most important factors behind the growing predicament of British capitalism in the post-war period.[86] More specifically, in terms of trade union organization, the movement's internal unity and its success in penetrating the workplace, the North Sea fares badly in comparison with the situation onshore. Thus a land-oriented safety regime which evolved within the context of the collectivism entailed by such 'defensive capacity' and was indeed, to a considerable extent predicated upon it, could only assume a somewhat alien appearance in the offshore setting.

As we have already seen, the philosophy endorsed by the Robens Committee in 1972 had placed great store by the development of a self-regulating system within which workpeople, as well as management, would 'participate fully in the making and monitoring of arrangements for safety and health at their place of work'.[87] This objective was subsequently built into the Health and Safety at Work Act under the auspices of which the Commission and Executive came into existence. Moreover, while the original statute allowed for the promulgation of regulations which would provide, in prescribed cases, either for the appointment of safety representatives by recognized trade unions or by their election among the workforce as a whole,[88] the latter alternative was subsequently removed by the Employment Protection Act of 1975.[89] In consequence, even by the time that the requisite regulations became operative with regard to onshore industry, the projected system of worker participation had become fairly firmly harnessed to the established machinery of collective bargaining.[90] By extension, the provisions whereby, again in prescribed cases, a duty could be imposed upon employers to establish safety committees 'if requested to do so by the safety representatives' became similarly bound up with trade union activity.[91]

While the administration of onshore safety was developing, albeit slowly, along these collectivist lines, the adoption of a similar approach to the safety of Continental Shelf operations was to prove difficult, not least because of the problems encountered in establishing the collectivist infrastructure itself. Despite the efforts of individual unions and the activities of the Inter-Union Offshore Oil Committee (IUOOC), an organization comprising eight trade unions with an interest in the offshore sector, progress towards unionization of the workforce on any substantial scale has been both contentious and slow. Union recognition had only been achieved on six out of twenty-eight rigs by the end of 1977,[92] and although some notable progress has subsequently been made in this respect by unions such as the National Union of Seamen (NUS), one informed observer still estimates the overall extent of unionization among offshore workers at less that 20 per cent in 1980.[93] While others suggest that such a figure is a substantial underestimate all are agreed that the process of developing union organization within the offshore industry has been particularly difficult.

Although detailed analysis of the factors behind the fraught history of trade unionism in the British sector of the North Sea lies beyond the scope of the present study, some brief discussion of this issue is necessary if the acrimonious implications spawned by the extension of the Health and Safety at Work Act to the Continental Shelf are to be fully appreciated. Here, right away, it must be sressed that the picture is not entirely one-sided. As we shall see, the industry certainly seems to be by no means enamoured of the robust processes involved in collective bargaining with organized labour, and this has undoubtedly played no small part in creating the difficulties which have been encountered. But this is not the whole story. According to some accounts, for example, the trade-union movement itself was somewhat tardy in developing anything approaching a co-ordinated approach to offshore unionization, and indeed it was not until 1974 that an action committee, comprising shop stewards from the shipbuilding industry, persuaded the Scottish Trades Union Congress to establish the IUOOC. Described by one participant as 'no small feat', even this step did not immediately guarantee the movement's future internal unity with regard to the North Sea, as became apparent when the NUS, probably the most active union offshore, capitalized upon its traditional position *vis-à-vis* the shipping industry and concluded deals outwith the local framework and the IUOOC in connection with members on semi-submersibles. At that point, it is said, the latter body almost split up. In more general terms, it has been argued that the British trade unions' fragmentary approach to unionization on the basis of traditional and traditonally demarcated skills bears little relationship to the close-knit and sometimes novel occupational tasks to be performed on offshore installations.[94]

On top of the problems stemming from the uneasy internal co-ordination of the trade union movement, unionization has also been hampered by other factors associated with the offshore workforce itself. Union officials readily concede, for example, that the labour force in question is not by nature one that is easily persuaded into union activity, partly because of apparently high wage rates, partly because of its itinerant nature – 'the same type of guy who built the Hydro Board scheme in the Highlands, the itinerant workers, the guy who will pack a bag and prepare to move anywhere' – and partly too because of the disillusionment with industrial unrest directly or indirectly experienced onshore. As the following extracts from interviews with union officials show, there is more to the problem of offshore organization than company hostility:

> But, you see, these guys out there are getting pretty well paid. . . . So it's big money and these guys have never seen that sort of money before; they've probably been used to £65, £70, £80 a week, even with overtime, so it's not altogether surprising that they are sort of saying, 'Well, we can't really complain; we've got a lot, plenty of money.' (Association of Scientific, Technical and Managerial Staffs [ASTMS] official, referring to production platform employees)

> Then, to be fair, we have had lads out on the rigs who have been in car plants or that, and got fed up with the continual unrest and that, and say, 'I'll cut myself clear', and [have] gone out to the rigs. Once more, you have got to look at the make-up of a man that goes out to the rigs. He usually comes from a high-unemployment area, usually an industrial area, or he comes from a country area [and] normally would emigrate to Australia, New Zealand or England for work. . . . That type of labour from the country is usually non-union. (NUS branch secretary)

> It's been an up-hill job. You know, people saying, 'Well, we've been so fed up with this other industrial situation onshore', and so the two things are combined in one: reluctance; not sure how the company would take it if they joined a trade union; and two, that if they did, they might be dragged into some dispute onshore. . . . (IUOOC officer)

> the employees on the platform have their own particular point of view about whether they should be a member . . . and some of them have rather outrageous points of view in regard to the trade union movement. In recent years the trade union movement hasn't had a very good press. One could say trade union movements never had a good press, depending on your point of view, but there have been certain situations onshore, related to certain industries, which have not projected the best possible image of the trade union movement. (District secretary, Amalgamated Union of Engineering Workers[AUEW])

Having conceded these difficulties, at least some of them self-inflicted, it must also be said that the overwhelming consensus of opinion among the union men interviewed in the course of this research was that the principal obstacle to effective offshore organization has been the attitude of the companies. Generally, it was suggested, there is no doubt that they would prefer a non-unionized workforce, a preference about which the Americans, in particular, were credited with a refreshing degree of honesty and frankness. Less starkly, even companies to which previous accounts have attributed a tolerant attitude[95] were described by several officials as being as anti-union and as obstructive as they could be without openly offering active opposition. According to the secretary of the IUOOC, 'They certainly give the impression of how willing they are, [but] I am quite sure they do everything to defeat that object.' In common with other companies, their basic approach was dubbed 'paternalistic' on more than one occasion. 'These are our children, we will look after them' was the impression reported by one member of the former Offshore Action Committee, while an AUEW official offered a staccato rendition of what he saw as the paternalistic approach common in the industry: 'You don't need a trade union; you're working offshore; we've got your interests at heart; this is a new industry; you're a big boy and you don't need the trade union to hold your hand; and we're looking after your interests.' Sometimes too 'looking after your interests' can apparently assume a very concrete form when the question of unionization is at issue:

> they will go to great lengths to try and discourage people to join unions, you know. They will, in fact, say to us that if it is the wish of oil people to become members, they will respect that wish. But, you know, for example . . . we established at least 75 per cent membership on the Piper platform; but before we could go to the Claymore platform, they in fact introduced a very, very favourable settlement which certainly did the trick with people on Claymore. Because in spite of the fact that some of them were on the Piper platform when we discussed joining the union and were quite keen and took forms . . . those very same people were saying, 'Well, we don't see there's any real need for the trade unions. The company has just given us. . . .' Well, in fact, they gave £10 a day. . . . (IUOOC secretary)

As with the carrot, so too on occasions, allegedly, the stick. ASTMS, for example, claims that its efforts to unionize production workers in one field have been impeded by moves to place relevant employees on hourly pay, thereby possibly jeopardizing their membership of that particular union and obviously barring them from the benefits which accrue to monthly paid staff. In another instance, an assistant crane driver was reported to have been dismissed, ostensibly for the unsafe operation of a crane, only a few days after he had been elected its representative by a newly formed trade union branch on a drilling rig. A progress report supplied by a union-appointed safety representative expressed similar misgivings about the implications of participating in union activity and claimed that verbal warnings had been given, 'with no witnesses around', that 'I was getting in too deep'. 'As I will put it,' the report concluded, 'veiled threats – quote – 'someone else got too involved in unions and he had to go'.' While the author of this particular document showed no signs of being put off by this reaction, such is not necessarily the case with many other workers, even with regard to the less active role of ordinary union member:

> Now on the X it was very significant. The lads said, 'Well, leave the rest to us. You've no problem here. It's just once we get it off the ground, we will make them members one at a time, build them up.' And they did it this way. And as long as a man knew that you weren't telling the company, he would join the union. So that rig has got 85 per cent, 95 per cent membership. Once we get that amount, it's OK; but individually they are frightened, and sometimes they project this big, hairy-chested, hard-case rig worker, and he is a bit embarrassed to say that he is frightened for his job because he is in the union. (NUS officer)

Explicit or implied, threats of dismissal are surely a powerful disincentive to involvement in offshore union activity and serve as a reminder that, despite all the rhetoric of advanced technology and frontier breaking, the North Sea oil industry may again in some respects be redolent more of the first half of the nineteenth century than of the latter part of the twentieth. Such tactics do not, however, completely exhaust the catalogue of methods allegedly employed to minimize the trade union presence. As for inspectors, so too for union officials: access to installations is ultimately under the logistical control of the operators, and several interviewees were less than reticent about the difficulties which could be created in this respect. According to one official, companies will also sometimes play a kind of numbers game, arguing that at least 50 per cent of an entire crew, 'including the rig master and all the professionals as well', should be union members before acceding to requests for recognition. More recently, a figure of 80 per cent has been reported as the requirement laid down by the companies.[96] In a similar context, I was told of one instance in which a company trouble-shooter claimed to have talked more than half of thirty-five men who had previously said they wanted to join the union out of that intention. On a more formal plane, it was claimed that employers make use of the existence of the IUOOC when approached by individual unions, and attempt to deflect them by asserting that all recognition issues must come through that body. It was even suggested that on some installations no opportunity is lost to expose the 'captive audience' to newspaper and magazine articles 'about how the unions are ruining the country'. All in all, union accounts of the difficulties which they have encountered offshore are very much in line with the forthright views expressed by the dissenting members of the Burgoyne Committee:

> Meanwhile in the UK sector the position is really quite scandalous. In spite of UKOOA [UK Offshore Operators' Association] understandings with the Inter-Union Offshore Oil Committee over trade union rights to access and recognition offshore, a battery of devices are used to minimize the trade union presence amongst permanent platform employees involved in production. To date there is only one collective bargaining agreement offshore, covering gas platform staff in the southern sector, and that was only secured after several years of resistance. Elsewhere pressure, intimidation, temporary bribes and other means are used to hold back the development of bona fide trade union organization and recognition.[97]

Needless to say, such charges are resolutely denied by the industry, which, in the main, publicly asserts its neutrality on the issue of unionization. In private, however, some of the oil men spoken to in the course of the project made little secret of their lack of enthusiasm for union involvement. A Texan rig superintendent, for example, confessed that he had 'never held with unions', their principal function being to secure 'more money for less work'. In similar vein, the general manager of one independent operator suggested that unions were 'very dangerous' because 'you can't have demarcation disputes on a rig', a reasoned, if not entirely reasonable, argument which he subsequently expanded by adding, 'If they don't want to work, they can get their ass off our rig.' No less forthright were the comments of the operator's representative on one drilling rig, who explained that union organization would be 'impossible due to the nature of the work' and detrimental because 'unions alienate employee/employer relationships in every industry.' More circumspectly, the personnel manager of another company – previously accorded some publicity for having reportedly gone through 18,000 British workers to find the right 350[98] – expounded the orthodox view that unions are unnecessary, since pay and conditions are probably much better than the men could expect onshore.

While such comments are only fragments of evidence encountered in the context of research which was not orientated to the investigation of labour relations as a major issue, they do little to allay the suspicion that there is considerable substance to union complaints. Moreover, when it comes to the more specific question of union participation in safety matters, there is no reason to suppose that this possibility would be greeted with any greater enthusiasm. For one thing, quite a few of the major companies and some of the smaller ones already operate a system of safety committees with representatives appointed from, or selected by, the workforce as a whole, and can therefore claim, with some plausibility, that formalization through union channels would be both unnecessary and possibly less representative. In the Forties field, for example, BP maintains a series of such committees as part of a policy of 'involving everybody in safety', and ensures that 'the men are elected from their little group and that there is a representative from each group at every meeting'. On a broader front, when worker participants at the 1977 International Labour Organization's meeting of experts on offshore safety pressed for recognition of 'the importance of workers' participation and trade union rights in safety and health matters', the employers' representatives, including officials from Shell, BP and Exxon, endorsed the first suggestion but not the second:

> The employer experts and several Government experts expressed concern that the meeting should avoid any statements which might limit the opportunities for all workers to involve themselves in safety and health matters, irrespective of whether or not they were represented by trade unions.[99]

From the active British trade unionist's point of view, such an appeal to a broader constituency, even in the restricted field of safety, represents something of an Achilles heel with which he is not entirely unfamiliar in other areas of collective bargaining. In the offshore context, however, it constitutes a special problem because it threatens to undermine effective union authority even before it has been established on any substantial scale. 'Castrated shop stewards' was the somewhat deprecating, if apologetically offered, description given by one union man, who went on to pinpoint the lack of 'real authority' as the system's major disadvantage and to draw parallels with the experience of the German labour movement and current Liberal Party thinking in Britain. To another, any scheme of this kind was bound to fail 'unless you've got the men organized in such a way that they can say, "Sorry, unless you do something about this, we're not prepared to continue in these conditions – that's it"'. The suspicion that management-instigated committees might not be sufficient to carry safety recommendations through to a satisfactory conclusion was also clearly one of the worries on the mind of the union safety representative mentioned above:

> On going aboard the rig, the OIM [Offshore Installation Manager] was informed by the convener of the appointment of a [union-specified] safety representative. We were met by what we think to be resistance to the idea of the union being involved in health and safety; or maybe they were frightened by the thought of it.
>
> The OIM then asked us what was wrong with the present system they used, to which we said nothing was wrong with it from their side, but it did not present a true view from the members' side. If they [the management] felt some item brought up did not suit them, or was too costly, or could upset someone sat behind a desk ashore, the item would be left out of the minutes. . . . The Safety Committee was hand-picked by rig management with no regard to the men, for example, the crane driver, driller, chief steward, bosun and others all being in charge of men.

From the foregoing, it will be apparent that the terrain into which the HSC and the HSE were drawn by the extension of the Health and Safety at Work Act to the North Sea was extremely treacherous. As we saw at the beginning of this section, the post-Robens developments in the administration of industrial safety onshore assumed an increasingly collectivist nature, in keeping with the established structure of labour relations, and it is within this framework that the Commission and its Excecutive seek to promote the philosophy of worker participation in safety affairs. Offshore, however, the requisite framework was largely absent, and a commitment to such an approach therefore not only raised practical difficulties, but also tended to become conflated with the broader issues involved in the struggle over unionization. Moreover, there is some evidence to suggest that the leading parliamentary advocates of the Act's extension were indeed motivated by the belief that 'if the companies were forced to set up committees in which trade unions were represented, that would mean they would have to recognize trade unions.'[100] John Prescott, Labour MP for Kingston-upon-Hull, East, is said to have been particularly persuaded by this argument, not least because of his scepticism about the possibilities of using the Employment Protection Act of 1975[101] to promote union organization on offshore installations. More generally, according to the head of Operations and Safety at the Department of Energy during the time that these developments were taking place, the pressure for extension of the Health and Safety at Work Act emanated from the Labour back benches as part of an attempt to unionize the industry.

In some respects, the approach adopted by Mr Prescott and his colleagues may have been ill-founded. Without too much difficulty, it would seem, a provision allowing for extension offshore was successfully inserted into the Employment Protection Act – 'It took a half-hour's conversation with Michael Foot and then a letter to him afterwards, and dash it, clause 113, I think it was, appeared.'[102] Less than a year later, moreover, the powers thus assumed were put to use in the Employment Protection (Offshore Employment) Order of 1976,[103] and the advocates of a strategy based on this body of legislation could claim a fair measure of success. Equally, they could assert, with some justification, that proponents of the alternative approach through the Health and Safety at Work Act had failed to appreciate that recognition is a prerequisite of representation and not vice versa. As one union official explained:

> if anyone thought that the Health and Safety at Work Act could be used as a vehicle for gaining trade union recognition offshore, then they were completely devoid of knowledge or understanding of what collective bargaining and the role of the trade union movement is all about. Because there is no employer that would be prepared, certainly in the offshore industry, to grant recognition on the basis that we have now got a Health and Safety at Work Act which has been extended to the offshore industry. . . . It just does not work in that fashion. Everything hinges, if we are talking about the Health and Safety at Work Act, everything hinges on Section 2(4), where, quite clearly the legislation refers to 'recognized trade unions'.

On the other hand, and while the above interpretation is legally correct, the parliamentary lobby for the extension of the Health and Safety at Work Act may have been shrewder than its critics allowed. Despite the legislative advances on the employment protection front, progress towards effective offshore union organization was not, as we have seen, particularly easy or rapid thereafter. Nor, notwithstanding its legislative commitment to the promotion of collective bargaining,[104] was it by any means certain that the government of the day would be prepared to exert additional pressure on offshore employers if they should prove recalcitrant. As Arnold remarked in this context, 'The government, whatever the Labour Party's ties with the trade union movement may be, has been more interested in getting the oil flowing fast and so has not wanted to promote more clashes with the companies than necessary.'[105] Thus there might be much said for a flanking

Oil, Safety and Internal Politics

approach that would generate pressure within the bureaucracy itself as a result of institutional objectives which, although not primarily concerned with collective bargaining, were nonetheless inextricably bound up with it.

With their commitment to operating within a collectivist framework onshore and to the development of harmonized regulations across industry as a whole, it was almost inevitable that, as a by-product of their health and safety functions, the HSC and the HSE would generate just such pressure in the context of their new-found responsibility for policy with regard to occupational safety offshore. More precisely, the contrast between their approach and that of the Department of Energy could scarcely fail to keep the issue of union organization alive, whatever the actual situation within the industry. Thus, for example, the Executive made a practice of consulting unions (both within and outwith the Oil Industry Advisory Committee) in connection with the formulation of regulations, and this in spite of the comparative dearth of union recognition offshore. Not suprisingly, the Trades Union Congress viewed this approach as being 'considerably superior to the procedures laid down or followed by the Department of Energy', procedures which it heavily, if not altogether accurately, implied comprised 'private on-going consultations purely between the companies concerned and Government Departments and agencies'.[106] Closer to the workplace, Section 28(8) of the Health and Safety at Work Act stipulates that inspectors shall furnish workpeople or their representatives with certain factual information about the premises and about their own actions relating thereto 'in circumstances in which it is necessary to do so for the purpose of assisting in keeping persons (or the representatives of persons) employed at any premises adequately informed about matters affecting their health, safety and welfare'. According to the union members of the Burgoyne Committee, the PED did not consult at all with such safety committees as existed offshore[107] – 'PED told us they have never needed to meet safety committees or safety representatives offshore, or advise them of their visits offshore.'[108] Certainly, too, none of the ordinary offshore workers spoken to or interviewed in the course of this research had ever had any dealings with such inspectors.

But the principal issue around which the unionization implications of HSC/HSE involvement came to revolve was the vexed question of safety representatives and committees themselves. Here, as Burgoyne points out, the Department of Energy's legislation includes no such concept,[109] while, as we have seen, this structured mode of participation through union-appointed representatives is central to the thinking behind the Health and Safety at Work Act. From the outset, however, the two agencies clearly had it in mind to extend the relevant regulations to the Continental Shelf once they had become operative on land. Indeed, even before the Act itself had been applied offshore, the Commission's chairman was talking of this step as one of the three things that would be considered immediately when the Order in Council went through.[110] Such optimism was not, however, to be justified in practice, the issue dragging on and on over the three years following the extension of the Act. Nor is there much doubt about the reasons for the delay. Time and again in subsequent interviews with HSE officials, I was told that the stumbling-block was the relative lack of union presence and recognition on installations. Within the Oil Industry Advisory Committee, UK Offshore Operators' Association objections and the problems associated with the frequent existence of multi-employers on installations held up agreement until November 1979,[111] and only thereafter was the Executive instructed to go ahead with drafting the necessary regulations. In complying with this request, the principle of optimal harmonization with onshore regulations was apparently adhered to once again, and, as with others of the HSE's regulatory plans which might involve greater union participation offshore,[112] yet more difficulties arose. By March of the following year, a further slippage in the timetable was being reported, the main problem again revolving around trade union recognition and the appropriate Acts.[113]

In the meantime, however, other things had also changed. A new Government had completed its first year in office and was moving into high gear with its monetarist policies and its plans to curb trade union power, neither of which augured well for the future of the HSE as a whole or for the extension of union-oriented regulations in particular. Moreover, the Burgoyne Committee had now reported and, apart from placing a large question mark against continued HSC/HSE involvement in offshore safety – a query which concomitantly, of course, cast doubts upon the future role of the Oil Industry Advisory Committee – had presented specific recommendations which were extremely ambiguous with regard to the way in which a revamped system for administering offshore safety might intersect with the structure of the industry's labour relations. Thus, while the participation of recognized trade unions in developing safety arrangements was applauded and the Department of Energy enjoined to provide for the requisite tripartite consultations,[114] the proposals regarding the participation of employees at the workplace itself were much more ambivalent. The fact that Section 2(5) of the Health and Safety at Work Act had been repealed, the appointment of safety representatives thus reverting to recognized trade unions, appears to have been overlooked[115] or discounted, and instead it was suggested that each installation should have a safety committee, 'the members of which are elected, appointed or co-opted to represent those employed for the time being on the installation'.[116] No less important, the Burgoyne Committee did not consider it essential that these and other principles concerning safety management should be embodied in mandatory regulations.[117] Hence, even while the HSE was struggling towards this objective, it was already being suggested that the effort was unnecessary. Against such a background, it is not suprising that progress during the spring and summer of 1980 continued to be slow, if not non-existent.

As with the other implications of HSC/HSE participation in the administration of offshore safety, then, so its possible impact upon labour relations became part of the internal politics of the state. Let me repeat: this is not to say that the extension of the Health and Safety at Work Act *per se* was imbued with such direct instrumental potentialities for union organization. At one level, the point is that this is how the step was seen by some of its proponents and, no less important, how their activities were interpreted by others. It would not be the first time in legislative history that an issue has assumed a measure of symbolic significance quite independent of what might actually be achieved in practice.[118] More crucially, my argument is that the letter of the law extant did not exhaust the contentious implications of HSC/HSE involvement in this context. Wedded to a collectivist and harmonized operational philosophy developed onshore, these bodies were unlikely to be easily infected with bureaucratic schizophrenia with regard to the North Sea. In consequence, their participation – particularly in a policy role – was bound to raise recurrent problems upon which the prevailing structure of offshore labour relations would have a direct bearing. The more they attempted to fulfil their allotted function, the more glaring became the gap between the position of unions within the offshore oil industry and their role in British industry as a whole. Equally, and even if the Health and Safety at Work Act could not insist upon it, the way forward under HSC/HSE auspices could best be facilitated by increased union participation at all levels. In contrast, the Department of Energy's approach would not require (and, one suspects, would not favour) such a development,[119] and the institutional battle lines on this issue thus became coterminous with those along which different parts of the state's bureaucratic apparatus were marshalled on others. Once again, there was much more to the politics of intra-state conflict than mere institutional pluralism and bureaucratic rivalry. What was at stake, by implication, was the extent to which an industry new to Britian would remain outside the parameters of the traditional processes of collective bargaining through which the British labour movement has, more successfully than most, exerted its defensive strength.

Notes and References

1. See, for example, Nicos Poulantzas, *Classes in Contemporary Capitalism*, London, Verso, 1978, p. 163; and Joachim Hirch, 'The State Apparatus and Social Reproduction', in J. Holloway and S. Piccioto (eds.), *State and Capital*, London, Edward Arnold, 1978, p. 100. For a good analysis of recent state theory, see B. Jessop, 'Recent Theories of the Capitalist State', *Cambridge Journal of Economics*, vol. 1, 1977 pp. 353ff.
2. See, in particular, J. Habermas, *Legitimation Crisis*, London, Heinemann Educational Books, 1976, pt II, ch. 5.
3. R. Miliband, *Marxism and Politics*, Oxford, Oxford University Press, 1977, p. 96.
4. J. Scott, *Corporations, Classes and Capitalism*, London, Hutchinson, 1979, p. 153.
5. In adopting this approach I am much encouraged by the perspective taken by James Petras in his work on Venezuelan oil. For a particularly trenchant criticism of abstract Marxist approaches, see the introduction to his book, *The Nationalisation of Venezuelan Oil*, New York, Praeger, 1977.
6. 1894 57 & 58 Vict., c. 60; 1970 c. 36; 1974 c. 43. The latter enactment provided the basis for the Merchant Shipping (Diving Operations) Regulations, 1975, SI 116, as amended by the Merchant Shipping (Diving Operations) (Amendment) Regulations, 1975, SI 2062.
7. A detailed description of the distribution of responsibilities between Departments from 1977 onwards is given in the Department of Trade's guide, *Marine Activities*, London, HMSO, 1977.
8. J. Kitchen, *Labour Law and Offshore Oil*, London, Croom Helm, 1977, p. 50.
9. *Offshore Safety* (the Burgoyne Report), Cmmd. 7841, London, HMSO, 1980, p. 14.
10. Department of Trade, *Marine Activities*, p. 26.
11. *Parliamentary Debates* (Commons), vol. 688, 1963–4, 28 January 1964, col. 234. For some discussion of the problems created in this context, see chapter 7 below.

12. *Safety and Health at Work* (the Robens Report), Cmnd. 5034, London, HMSO, 1972.
13. ibid., p. 7.
14. ibid., p. 35.
15. ibid., p. 12.
16. ibid., p. 319.
17. ibid., pp. 11–12.
18. ibid., p. 32.
19. C. 37.
20. ibid., S. 2(b).
21. ibid., S. 10.
22. *Parliamentary Debates* (Commons), vol. 871, 1974, 3 April 1974, col. 1287.
23. *Parliamentary Debates* (Commons), vol. 991, 1980–1, 6 November 1980, cols. 1530–1.
24. See above, p. 145.
25. Interview with HSC chairman, November 1976.
26. Robens Report, p. 35.
27. ibid.
28. S. 53(1). By default, as it were, the statute also acknowledged this possibility by Section 15(9) which stipulated that in the event of an Order being made under Section 84(3) health and safety regulations would not apply, *inter alia*, to offshore installations outside Great Britian 'except in so far as the regulations expressly so provide'.
29. *Department of Trade Memorandum on the Allocation of Responsibilities for Offshore Safety*, July 1976.
30. See below, p. 254.
31. See Burgoyne Report.
32. The best resumés of the reorganized framework for offshore safety can be found in the text of the Prime Minister's statement referred to above, in the Department of Trade's guide to marine responsibilities, *Marine Activities*, and in chapter 2 of the Burgoyne Report. The 'agency agreement' referred to was delayed for a considerable time and did not come into effect until 1 November 1978. One reason for the delay was that Sections 13 and 18 of the Health and Safety at Work Act created some ambiguity as to whether it was the Commission or the Executive that should negotiate such agreements.
33. SI 1977, 1232.
34. For details, see Burgoyne Report, p. 97.
35. For details, see the *Report from the Commission on Inquiry* (Uncontrolled Blow-Out on Bravo), 1977; unofficial translation supplied by Norwegian authorities.
36. The Burgoyne Committee.
37. 27 April 1980.
38. Burgoyne Report, p. 1.
39. ibid., p. 14.

40. ibid., p. 69.
41. See, for example, the note of dissent appended to the main Report, ibid., p. 58. Here it is pointed out that the Committee's recommendations impinged not only upon the role of the HSE, but also upon that of both the Department of Trade and the Civil Aviation Authority.
42. ibid., p. 3.
43. ibid., p. 13.
44. ibid., p. 58.
45. ibid., p. 15.
46. ibid.
47. ibid., p. 18.
48. Robens Report, p. 212.
49. ibid.
50. *Health and Safety Commission Report 1974–76*, London, HMSO, 1977, p. 4.
51. ibid., p. 35 and appendix 5.
52. ibid., p. 35.
53. Daintith, for example, points out that S. 3(1) of the Continental Shelf Act 1964, which extends the criminal law of the UK to an area of 500 metres around installations, seems to exceed its competence under the Convention in several respects, not least because Article 2.1 of the Convention grants sovereign rights only for the purpose of exploring and exploiting the natural resources of the Shelf. T. Daintith and G. Willoughby (eds.), *A Manual of United Kingdom Oil and Gas Law*, London, Oyez Publishing, 1977, p. 178.
54. *Time Out*, 16–22 November 1979. According to this report, the Attorney General wrote to the Secretary of State for Social Services saying that it was 'of the utmost importance that the Filby Case should if at all possible be settled as soon as possible'. He had spoken to the Insurance Commissioner, so the report of his letter claimed, and this official had stated his willingness to allow the appeal to be withdrawn 'provided Filby had advice of the best quality that an *ex gratia* payment would make him at least as well off as he would be if he won the case in the European Court'. Subsequent efforts by journalists to trace Filby have not proved successful, and the Social Security (Persons Abroad) Amendment Regulations of 1979 ensure that future embarrassment on this score at least can be avoided.
55. Daintith, in Daintith and Willoughby (eds.), *A Manual of United Kingdom Oil and Gas Law*, p. 217.
56. Burgoyne Report, p. 14.
57. Chairman's note, OSI(79)37.
58. *Health and Safety Commission Report 1974–76*, p. 11.
59. Burgoyne Report, p. 160.
60. This concern may seem paradoxical in the light of the earlier discussion of the slow pace at which regulations were made. The point is, however, that once the industry knows that regulations on a particular

score are inevitable, it wishes to know as rapidly as possible what is going to be required. The UK Offshore Operators' Association complaints in this context are as cited in the Burgoyne Report, p. 248.

61. 'PED's Comments on Occupational Health and Safety Work of the Petroleum Engineering Division – HSE's Views', October 1979. The terms 'Division' and 'Directorate' seem to be used interchangeably.
62. Burgoyne Report, pp. 229–30.
63. ibid., p. 197.
64. See above, chapter 5.
65. In interview, one HSE official went to considerable lengths to explain that the uniqueness of offshore process was frequently exaggerated. Many of them, he suggested, parallel processes in manufacturing industry, construction, mining and quarrying.
66. Burgoyne Report, pp. 16–17.
67. ibid., p. 229.
68. ibid., p. 140.
69. See above, chapter 5.
70. Burgoyne Report, p. 189.
71. 'PED's Comments', p. 3.
72. Robens Report, p. 35.
73. ibid., p. 32.
74. ibid., p. 33.
75. See above, chapter 5.
76. Burgoyne Report, p. 15.
77. *Current Account*, 29 April 1980.
78. See below, chapter 7.
79. Burgoyne Report, p. 16.
80. ibid.
81. See, for example, my papers: 'White-Collar Crime and the Institutionalisation of Ambiguity: the Case of the Early British Factory Acts', in G. Geis and E. Stotland (eds.), *White-Collar Crime*, Beverly Hills, Sage Publications, 1980, pp. 142–73; 'Some Sociological Aspects of Strict Liability and the Enforcement of Factory Legislation', *Modern Law Review*, vol. 33, no. 4, 1970, pp. 396–412; 'White Collar Crime and the Enforcement of Factory Legislation', *British Journal of Criminology*, vol. 10, 1970, pp. 383–406.
82. *Guardian*, 28 January 1980. More recently, the HSC has distinguished itself among the many bodies under threat from Government expenditure cuts by reportedly telling Ministers that they will have to take responsibility for deciding how such cuts should be put into operation, ibid., 26 July 1980. If these reports are correct, they bear out the interpretation of 'relative independence' being offered here.
83. 'Note of Dissent', Burgoyne Report, p. 58.
84. See above, p. 8.
85. Miliband, *Marxism and Politics*, p. 97.

86.   D. Purdy, 'British Capitalism Since the War', *Marxism Today*, vol. 20, 1976, pp. 270–318.
87.   See above, p. 191.
88.   Sections 2(4) and 2(5).
89.   1975, c. 71, S. 116 and schedule 15. The effect of these provisions was to repeal Section 2(5) of the Health and Safety at Work Act, which, at the instigation of the House of Lords, had been added to the original Bill in order to allow the election of representatives by the workforce as an alternative to trade union appointment. See B. Bercusson, *The Employment Protection Act, 1975*, London, Sweet & Maxwell, 1976.
90.   Partly as a result of the above duplication, the relevant onshore regulations did not come into effect until October 1978.
91.   S. 2(7) as amended by the Employment Protection Act, 1975, Schedule 15.
92.   Guy Arnold, *Britain's Oil*, London, Hamish Hamilton, 1978, p. 293.
93.   P. Watson, 'Trade Union Activity in the Offshore Oil Industry', Ph.D. thesis, forthcoming, University of Dundee.
94.   ibid.
95.   Kitchen, *Labour Law and Offshore Oil*, p. 100.
96.   *Daily Telegraph*, 5 June 1980.
97.   Burgoyne Report, p. 62.
98.   Kitchen, *Labour Law and Offshore Oil*, p. 100.
99.   *Safety Problems in the Offshore Petroleum Industry*, Geneva, International Labour Organization, 1978, p. 98.
100. This is the view expressed by one heavily involved union official who was critical of the idea that even with the repeal of Section 2(5) of the Health and Safety at Work Act, its extension would necessarily promote unionization.
101. Whether this scepticism was based on doubts about the possibility of securing that Act's extension or upon misgivings about its efficacy even if it were has not been established.
102. This is, in fact, almost certainly an allusion to Section 127(2)(b) of the Act, which provided for extension to the Continental Shelf.
103. SI 1976, 766.
104. Under Section 1(2) of the Employment Protection Act 1975, the newly established Advisory, Conciliation and Arbitration Service was charged 'in particular' with the duty of encouraging the extension of collective bargaining.
105. Arnold, *Britain's Oil*, p. 294.
106. Burgoyne Report, p. 243. Kitchen, for example, points out (*Labour Law and Offshore Oil*, p. 103) that the Trades Union Congress was consulted about regulations made under the Mineral Workings (Offshore Installations) Act 1971, but union leaders remain sceptical, particularly of the procedure whereby discussions on safety take place between unions and UK Offshore Operators' Association 'purely so that they can go to the Department of Energy and say we are talking to

230 — Oil, Safety and Internal Politics

the trade unions'. The point seems to be that consultation is institutionalized within the Oil Industry Advisory Committee as part of HSE policy.

107. Burgoyne Report, p. 63.
108. ibid., p. 60.
109. ibid., p. 41.
110. November 1976.
111. The 'Note of Dissent', Burgoyne Report, p. 63, attributes a delay of about twelve months to the UK Offshore Operators' Association's objections, although it does not say what these were.
112. Proposed regulations on the training of divers are reported to have encountered similar problems.
113. Other problematic areas of interface include the Trade Union and Labour Relations Act 1974, c. 52, and the Merchant Shipping Act 1974, c. 43.
114. Burgoyne Report, p. 17. Proposed regulations under the Petroleum and Submarine Pipelines Act 1975, c. 74, have subsequently been submitted to the Scottish Trades Union Congress by the Department of Energy.
115. Burgoyne Report, p. 41. Paragraph 5.94 attributes to the Health and Safety at Work Act provisions for 'the appointment or election of safety representatives to represent employees in consultations with the employer on safety matters'.
116. Burgoyne Report, p. 42.
117. ibid.
118. See, for example, my paper, 'Symbolic and Instrumental Dimensions of Early Factory Legislation', in R. Hood (ed.), *Crime, Criminology and Public Policy*, London, Heinemann, 1974, pp. 107–38.
119. In interview, for example, one PED inspector gave it as his opinion that, quite apart from the dangerous implications of a stoppage offshore, the further unions could be kept away from the industry, the better it would be for all concerned.

7

Compromise, Chaos and Control:
the Offshore Safety Regime
in Operation

*... to put it another way, the economy of illegalities was
restructured with the development of capitalist society. The
illegality of property was separated from the illegality of rights.
This distinction represents a class opposition because, on the one
hand, the illegality that was to be most accessible to the lower
classes was that of property ... and because, on the other, the
bourgeoisie was to reserve to itself the illegality of right: the
possiblity of getting round its own regulations and its own laws, of
ensuring for itself an immense sector of economic circulation by a
skilful manipulation of gaps in the law — gaps that were foreseen by
its silences, or opened up by* de facto *tolerance.*
Michel Foucault,
Discipline and Punish

It's a mess. The whole thing's a mess!
Crown Office Official, 1978

Ever since Britain first underwent the throes of industrialization, rules
pertaining to the safety, health and welfare of industrial employees
have occupied a significant position within the broader economy of
illegalities. While the laws in question created criminal offences and
provided criminal sanctions, the enforcement of early factory legislation rapidly took on an educative, advisory and cautionary quality
which relegated prosecution to the status of last resort. And this it has
largely retained in the current era, when the corporation has sup-
planted the individual entrepreneur as the main employer of industrial
labour and the state has appropriated the function of law making in
the name of a universalized interest. Indeed, the history of this type of
legislation and of the acts and omissions which it covers has to be one
of what has been dubbed 'conventional crime' — crime that is accepted

by custom, is rarely thought of as 'real' crime and is only infrequently punished under the auspices of the criminal law. Arguably the paradigmatic example of offences known to criminologists as 'white-collar crime', violation of the rules imposed by or under statutes dealing with the safety, health and welfare of employees elicits largely non-penal strategies and rarely, if ever, figures in the rhetoric of the 'war against crime'. *De facto* tolerance has become institutionalized, and violation has indeed been sustained as an illegality of right.[1]

Far from being an exception to this general pattern the history of offshore safety regulations has become an extreme example. As we shall see below, the agency primarily involved in this field has only very infrequently had recourse to legal proceedings, and indeed the enforcement culture which it has developed has hitherto been one that does not accord much salience to legal norms as an organizing system of categories for purposes of routine operational processes. Moreover, while reluctance to employ legal sanctions does not mean that other strategies for securing compliance are eschewed, the overall picture seems to show tolerance of non-compliance being institutionalized at a fairly high level. Particularly apparent in relation to those routine aspects of occupational safety which were earlier seen to be the issue at stake in the bulk of accidents occurring in the British sector of the North Sea,[2] this institutionalized tolerance has become one of the distinctive characteristics of the offshore safety regime's operations. Thus, although the question of taking steps to ensure the safety of activities on the Continental Shelf first arose within the context of legislative moves to bring 'law and order' to the North Sea, this particular aspect of offshore operations can scarcely be described as having found a place in the annals of the recurrent law and order campaigns of recent years.

The first task to be undertaken in this chapter is that of mapping out the broad parameters of this institutionalized tolerance. Before this can be done, however, a prior issue to be addressed is the level of analysis at which such a discussion should be pitched. Clearly, for example, the issue of power differentials could be important in this context, since most of those who have written about conventional and 'white-collar' crime have alluded to this dimension of social structure as an important determining factor behind the low-key and ambivalent responses elicited by these types of offence. As Alvin Gouldner, one of the proponents of the view that 'the powerful can conventionalize their moral defaults', remarks in a more general context, 'The level at which moral default comes to be established is, in large part, determined by the relative power of the groups involved.'³ Certainly, too, it is the case that the powerful offender, unlike the run-of-the-mill criminal, can often mobilize highly effective resources to protect his moral and legal propriety in the event of action being taken against him in the courts.⁴ Closer to the subject at hand, the power of the body comprising the employers of industrial labour undoubtedly played no small part in creating a situation in which, from the very beginning of legislation pertaining to working conditions, putative and actual offenders could count upon a fair degree of deference being shown towards them by legislatures and enforcement agencies alike.⁵ In broader terms, the exercise of power – both in the past and in the present – has been seen as one of the crucial components of the process whereby principles of state action become institutionalized:

> Institutionalization is an unintended consequence of the attempts made by particular interests to influence the state. Over time, a structure is built up which has the 'bias' of past pressures built into it. This bias creates a predisposition for the state to respond in the future in the same way as it has in the past. Unsuccessful responses lead to failure and so are not institutionalized. State apparatuses are, therefore, formed through a complex adaptive process.⁶

In view of the corporate strength and the strategically advantageous position occupied by the interests involved in offshore exploration and exploitation, it would not be surprising to discover that they have been able to enjoy a considerable measure of success in asserting their power in all of the above contexts. Indeed, much of what has been said in previous chapters is consistent with the successful deployment of such influence, and there is no *a priori* reason to assume that law enforcement would prove to be an exception. Elsewhere, however, I have suggested that analysis of conventionalization, institutionalized tolerance and the rest has to go beyond what could all too easily become a crude portrayal of legislatures, enforcement agencies and courts simply succumbing to the undoubted strength of structurally powerful interests.⁷ To be sure, power is by no means irrelevant, and the marks of its application are indeed there to be seen in the complex adaptive processes of institutionalized action. But there is more to it than that. In addition, it has to be remembered that the front-line agencies involved in the enforcement of laws such as those under consideration here are operating under constraints imposed by the overarching and often contradictory imperatives stemming from forces which transcend the direct exercise of power by interest groups. More specifically, it is my view that their practices have to be analytically reconnected with a wider and more coherent sense of the totality within which they function than is encompassed by a diffuse vision of myriad interest groups and the uneven distribution of power. In the present context, one vital part of such a project would be to reconnect the institutionalized tolerance of offshore safety contraventions with the broader issues of political economy discussed in chapter 4 of this book.[8]

It is with such connections that much of the present chapter will by implication be concerned. In the first section, I will attempt to show that such links not only are plausible but have also been articulated by observers and, on occasion, by participants themselves. Actors do not always perform in complete ignorance of the constraints which shape their action. In analogous vein, the second part of the discussion will focus upon the importance of such links for an understanding of the confusion which came to surround the operation of offshore safety laws. At one level, it will be suggested, not all of the legal anomalies and jurisdictional gaps which have plagued law enforcement in the North Sea can simply be laid at the door of faulty draughtsmanship and understandable oversight – the loopholes which have indeed on occasion been skilfully manipulated often have a more substantial aetiology than inadvertence. On another plane, I will argue that no small measure of chaos was added to the problems of enforcement by the fact that the regulation of offshore safety was conceived largely within the broad context of 'British' interests. As a consequence, the distinctive law and legal administration of Scotland were to encounter major difficulties in implementing the law when the focus of offshore attention shifted to waters off the Scottish coast. Just as combined and uneven development on an international plane was advanced in chapter 4 as an organizing theme for understanding the political economy of speed, so it must not be forgotten that analogous links and disjunctions permeate the relationship between the component parts of the United Kingdom. In short, and for historical reasons which cannot be elucidated here, there was to be a pronounced disarticulation between Scotland's law and a safety regime created and operated in connection with Britain's oil. In a final section, we shall see that problems stemming from this source, along with those emanating from the other factors already mentioned, provided plenty of opportunity for 'getting around regulations', even when the ultimate sanction of criminal prosecution was invoked.

The Institutionalization of Tolerance

In chapters 4 and 5 I charted the combination of forces which engendered the speed of North Sea development, and I looked at some of the consequences in terms of both general relationships and the more specific relationship involved in the legal regulation of safety matters. Not surprisingly, however, these same forces also had an impact upon the practice of enforcement. Creating problems, predilections and preoccupations for enforcers, they helped to attenuate the role of law *vis-à-vis* the safety of offshore operations, and, in so doing, they made a substantial contribution to the institutionalization of tolerance. Since this impact was most apparent in the case of the Department of Energy, the body primarily involved in the enforcement side of North Sea safety regulations, it is on the activities of this organization that the present section will largely concentrate.

One immediate effect of the forces underpinning the rapid development of the Continental Shelf will already be apparent from what has been said at an earlier stage.[9] Because development outran the statutory response with regard to safety, for more than half a decade inspectors did not have any statutory regulations to enforce. As we have also noted above, moreover, even after the necessary steps were taken to rectify this situation in 1971, the relevant officials remained preoccupied with drafting rather than with administering regulations for some time, with the result that they are understandably modest about imbuing the product of their labours with the status of categorical legal imperatives.[10] Furthermore, while the problem of regulatory lacunae may have receded over recent years, this process has been neither particularly expeditious nor complete. Thus, for example, there are still no regulations covering the safe construction and operation of pipelines, although at the time of writing steps are being taken to plug this gap,[11] and up until 1979 the Department of Energy's legal advisers adhered to the view that accommodation platforms moored alongside fixed installations ('flotels', like the Alexander Kielland) did not constitute 'offshore installations' and were not therefore subject to control under the 1971 Act.[12] Similarly, and although the authorities were warned in the course of debates surrounding the latter enactment that the construction phase was particularly hazardous,[13] there are still no regulations covering occupational safety during this stage in the process of establishing an offshore installation. While the Health and Safety at Work Act does apply, no regulations have been promulgated under its auspices, and the regulations relating to operational safety, health and welfare and to emergency procedures – imposed under the Mineral Workings (Offshore Installations) Act of 1971 – only apply to 'established fixed offshore installations'.[14]

Depending on the context then, inspectors have not always been in the position of having statutory codes at hand to back up their suggestions or to nail down observed deficiencies. Even when they have been, however, this has not necessarily meant an end to their difficulties. Diffident about their regulations, they may find themselves dealing with companies which are brimful of confidence and which do not therefore instinctively lapse into a posture of deference before the invocation of legal norms. As the following extract from one inspection report shows, even when an inspector is firmly convinced about the correctness of his own position, his criticisms may not always go uncontested:

> The air supply for the purging air [for the Logging Cabin] was taken from immediately behind the cabin. This point of ingress is approximately 15 feet from the Shale Shaker door in one direction and a similar distance from the Mud Tank Room mushroom ventilator in the other direction. X's engineer was present and argued that this was acceptable under their [Certifying Authority] survey. In the writer's opinion this is not so as the purging fan is sucking from an area into which the mushroom vent is discharging. A direct flow of air could easily be established from one to the other. This point was not resolved to the satisfaction of both parties during the inspection.

Further evidence of company self-confidence emerged from interviews with senior safety personnel at BP. According to one official, for example, 'BP sets its own standards and has done as most oil companies have done through the years, because there has been no legislation covering oil operations when you are in a country, an underdeveloped country, and you are your own regulations.' In fairness, it must be added that the broader import of this comment was to underline the fact that the company's standards are usually higher than those laid down by British law, but at the same time, no secret was made of the fact that acquiescence might not always be uncontentious. Asked to assess the corporate reaction to an eventuality such as the imposition of a prohibition notice,[15] the same respondent replied, 'If from the company's viewpoint [it] was warranted, then of course we would endorse it.' Similarly, he explained that there had been problems in the early days, when the managers of the smaller, southern platforms had been comparatively junior and were 'very reluctant to have open discussions with the inspectors'. Instead, when the latter said ' "Do this, do that", they would say, "Yes sir, yes sir, yes sir," whereas there would probably be a perfectly good reason why one didn't do it that way'. But the problem had now been solved, partly by the more senior status of managers in the Northern Basin, and largely by sending out a senior and technically qualified safety man with the inspector. As a result, there would now be an on-the-spot dialogue to resolve the problem. The implications for the practice of enforcement (as, incidentally, for the relevance of the micro-sociological dimension to an understanding of how rule application is negotiated) were pointed out by another company official:

> What we often find is that if we don't have that stage and the inspector goes away and then writes to us about a number of points, he almost inevitably is forced to take a stand which he might not particularly want to see through in the end. But he's got it down in black and white, his boss is looking over his shoulder, and he feels that he's got to push this thing through.

BP, of course, is a British firm and, as such, might be expected to adjust fairly readily to the requirements of UK law, extensive involvement in other parts of the world notwithstanding. Again, as we saw above, however, another effect of the commitment to speed in the development of the North Sea was the heavy reliance upon foreign companies, and here the ease or willingness of adjustment cannnot be quite as comfortably assumed. After all, operations may be taking place several hundreds of miles offshore, and, niceties of jurisdictional demarcations apart, management may not take too kindly to the intrusion of British officialdom and its regulations. Speaking in the debate on the Second Reading of the 1971 Act, for example, one MP welcomed the fact that its provisions applied expressly to non-British owners, because 'foreign owners of rigs operating in British waters might well need a salutary reminder in advance that British law will apply to them in this matter.'[16] That such a reminder might be in order for some personnel further down the chain of command was apparent from the response which a barge captain gave in interview to questions about British jurisdiction. Whatever the barge's proximity to a UK installation, whatever the jurisdictional pretensions of the 1975 Petroleum and Submarine Pipelines Act, he strenuously insisted that his vessel came under American law. Whether inspectors, in this case the HSE, would even be furnished with transportation for purposes of enforcement was open to question:

> They say they have powers to board, but they would have to go through the company and virtually demand transport to the barge. The only way they will get on the barge is by helicopter, and it will not be allowed to land without permission from the control tower – or by the supply boat who would say they couldn't take the supply boat alongside, but they can. . . . Perhaps the company would refuse [transport], but if they are already complying to the American standards . . . there is no reason for the British inspector to come on board. In a way, it is none of his business unless he states that there is British personnel working on it. . . .

While this example may owe quite a lot of its residual intransigence to the protracted confusion which was allowed to surround pipelaying activities in the North Sea, there are still plenty of suspicions that the enforcement of British legislation on installations themselves can cause irritation. According to one union official interviewed in 1978, for example, most of the companies have American tie-ups even if they are not directly American, and 'They still think they are working in the Middle East or off the coast of Africa somewhere and [that] they can get away with anything.' 'To hell with the law. I'm the law out here' was the explicit attitude attributed to an American toolpusher by an erstwhile rig worker, while a British official was able to claim, again in 1978, that 'all too often, American oilmen look on the worker-safety standards which now supposedly apply, as "molly-coddling".'[17] More direct experience of such impatience was obtained during a visit to one semi-submersible, when the on-board representative of the US operators expressed his resentment at various aspects of British safety legislation. In particular, he suggested, the regulation which requires that the site of casualties involving loss of or danger to life must not be disturbed for three days or until an inspector has completed his investigation was an example of heavy-handed irrationality.[18] Such accidents almost invariably happen on the drill floor, he insisted, with the result that operations have to stop and thousands of dollars are lost. As for the victim, 'if he's dead, put him on a helicopter and send him ashore – let them meet the helicopter.' More generally, the attitude of this particular oil man to British safety regulations was summed up in a similarly business-like statement that warrants verbatim quotation: 'We break your fuckin' law every fuckin' day; if we didn't, you wouldn't have one fuckin' hole drilled in your fuckin' North Sea.'

Although such forthright recalcitrance is categorically not being advanced as typical of the position adopted by offshore management, foreign or indigenous, it does raise the inevitable question of what the inspectors can do when such attitudes materialize in the practice of offshore operations. How, too, can they counter the practical consequences of the more common situation in which legal intervention may be seen either as unnecessary pampering of the workforce or as a somewhat superfluous exercise in light of a company's in-house codes? Their first and most obvious problem is that of getting there, and, in one sense, the barge captain cited above was correct. Even for purposes of carrying out inspections or investigating accidents on installations, they have to rely on the companies for helicopter transportation. In the less confused jurisdictional context of installations as opposed to barges, however, the Department of Energy's inspectors are empowered to require the provision of transport at any reasonable time under the Offshore Installations (Inspectors and Casualties) Regulations of 1973,[19] and in the course of this project, no instance was encountered of such a request being refused. Normally, the inspector will give something like a week's notice of his intention to go offshore, and the operator will make the necessary arrangements for helicopter space and for accommodation on board.[20]

According to some of the people interviewed during the present research, this system contains vital flaws as far as the proper enforcement of British safety regulations is concerned. Spot-checks, it was pointed out on more than one occasion, are precluded, while offshore managements will obviously have time to take at least some remedial action in advance of a visit. In addition, the warning of an impending inspection will also alert them to whether the inspector in question is a specialist on the electrical, mechanical, petroleum engineering or structural side of matters, with the result that particular attention can be paid to those areas in which he is likely to take most interest. Recalling reports from some of his offshore members, one union official made the general point with characteristic acerbity:

> *Union official:* ... for some reason or other an inspector is coming out and there's all hell let loose. They are right busy little bodies getting everything spick and span for the inspectors. But obviously, if this is the case, the necessity here is for the inspector to go up to that airport and commandeer a seat on the helicopter for that rig, or alternatively....
>
> *Interviewer:* They can, of course, do that. They have the legal power to do it. It's just that they don't exercise it.
>
> *Union official:* They don't exercise it. What an inspector does is to get in touch with the company and say, "I'm going

> to visit your rig [and] I think I'll come out next
> week' in a very British type of manner. Well, of
> course, that's no good to those fellows out on the
> rig. The ultimate would be, of course, for our in-
> spectors to have a helicopter of their own, so they
> can pop out any time.

Although it is probably the case that the Department of Energy
could usefully adopt a logistically less deferential approach to inspection, too much can be made of this criticism. One point in mitigation,
again not always unrelated to the speed of offshore development, is
that sleeping accommodation is often at a premium, particularly in the
early stages of construction or when, as has happened in many cases,
construction work has carried over into drilling or production-related
operations. Equally, regulations covering the safety of helicopter operations would mean that *some* prior warning of a visit would be
received in any event, since no helicopter is allowed to land on a
manned offshore installation without radio communication having
been established before take-off.[21] While spot-checks may not be
possible in the full sense, furthermore, the Inspectorate can plausibly
claim that this deficiency is offset somewhat by the fact that, on
average, each installation is visited every three to four months, an
inspection rate which allows deficiencies previously recorded to be
followed up and one which, indeed, is said to have attracted criticisms
of over-inspection from the Health and Safety Executive.[22] As for
remedial measures being taken in advance, the former head of Operations and Safety in the Department had no qualms: 'So they put their
known problems right *before* our arrival – a bad thing?'

Allowing that the logistical problems of inspection have sometimes
been overemphasized, what happens once an inspector has boarded an
installation? Although the Department has been slow to establish
standardized procedures, the normal practice seems to be for the
inspector to carry out checks on items of plant and equipment falling
within the area of his own particular expertise and, in addition, to
cover the more obvious matters of general concern. Engines on life-
saving capsules, crash equipment on helidecks, logbook maintenance
and sick bays were all instanced as items typically covered in this more
general part of the inspection. Any matters requiring attention are
discussed with the company personnel accompanying the inspector,
and a note itemizing the issues in question is then left with the installation's manager. A note may be entered in the installation's logbook,
and in either case a copy is taken back to shore and a fuller report of the

visit is then submitted to the Department's headquarters in London. Direct correspondence between London and the company may follow, or a further visit may be arranged ahead of the normal schedule if there is anything giving rise to particularly serious concern.

Although it has not been possible to gain access to a representative sample of Department of Energy inspection reports in connection with the enforcement of offshore safety regulations, evidence drawn largely, though not exclusively, from other sources permits the broad pattern of institutionalized tolerance which has emerged to be mapped out. To begin with, it does seem to be the case that the inspection process outlined above tends to revolve around a central preoccupation with major hazards. At one point in the research, for example, the then head of Operations and Safety in the Department admitted to being unsure as to what occupational safety really was, the whole field of offshore safety being, as far as he was concerned, subsumed under the heading of either 'structural' or 'production' safety. For good measure, he added, 'The Health and Safety Executive fuss about small details which barely touch on the major hazards offshore.' The Division's director was similarly less than enthusiastic about what he saw as attempts to shift the focus of attention more towards the occupational side of things. 'The fear in PED', he observed, 'is that concentration of the 'occupational safety'' of some 20,000 men will interfere with the control of the very real and potentially very much greater hazards they face from structural failure, tempest, fire or blowout.' Significantly, too, this attitude seems to have been a source of irritation to some safety personnel within the industry itself. Indeed, according to the safety manager for one well-known oil company, the inspectors' approach actually created safety problems because of its cursoriness and lack of attention to the detail of more mundane issues. As a result, he contended, the incentive to improve this aspect of offshore safety was not being backed up, and a lot of the effort put into making the installation 'shipshape' appeared to be wasted. Such views serve to underline the doubts expressed by an HSE official on the same score:

> Quite often they get the big things right – the big things are politically more dangerous for an enforcing authority and, sure, you've got to get that right as well. But I sometimes think we get it right because the current political implications of not getting it right would be far worse. But, you know, you've got to get the little things right. That's where you make the biggest impact on accidents, and how the Department of Energy are on that sort of thing I have no idea at all. I'm really not very sure of what the background of these inspectors are, so there again, it's something they can learn. . . .

Once again, it must be emphasized here that the PED's preoccupation with major hazards is quite understandable when viewed against the background of North Sea developments, which have constantly been in danger of outstripping not only the legislative response but also the accumulation of the expertise and knowledge requisite to the prevention and handling of such contingencies.[23] Fairness also justifies repetition of the point that, to date, no major catastrophe has occurred in the British sector since the loss of the Sea Gem in 1965 (but see p. 5). This said, however, the almost inevitable corollary seems to have been that toleration of more mundane deficiencies in occupational safety standards became established at a relatively high level by comparison with what would be accepted onshore. While the Department of Energy was heading off disaster with some evident measure of success, the 'little things' continued to take their insidious toll of the offshore workforce, and the role of law in the protracted business of raising the standard of routine, everyday operations was undervalued.

This imbalance in the emphasis placed upon different aspects of safety is not, however, the only way in which the broader forces underpinning North Sea developments permeated the practice of inspection and enforcement. Of possibly even greater importance is the fact that, as we have already seen, the offshore enterprise was, from the outset, almost predicated upon the need to maintain as rapid progress as possible towards the production of indigenous supplies and, ultimately, towards the goal of self-sufficiency. In such circumstances, it would be more than slightly optimistic to expect the enforcement of safety regulations to remain entirely impervious to the overriding imperatives generated by such a commitment, and indeed, as was pointed out in chapter 6, there is no dearth of outside observers who feel that the Department of Energy is caught on the horns of a sizeable dilemma in this respect. As the following case history of one of the most important North Sea platforms shows, moreover, such constraints have not only asserted themselves on occasion, but have also found their way into conscious expression at the hands of the inspectors involved.

In late 1974, this particular installation was inspected while it was still, in effect, little more than a maritime construction site, and it was noted that the intention was to house 122 persons in temporary 'portacabins', between six and ten to a cabin, during the construction phase. However, no Certificate of Fitness under the Construction and Survey Regulations, 1974 was required at this point because the absence of any oil-related function on an installation at this stage of development has been interpreted by the Department as exempting it from this requirement and because, anyway, the deadline fixed for certification by the Regulations was August 1975.[24] Such a certificate, with its prescription of appropriate accommodation standards, escape routes and protection from hazards such as fire, would therefore only become necessary when drilling commenced, a stage that was reached in June 1975, some eight or nine weeks before the deadline was due to expire. During an inspection carried out at this point, it was noted, among other things, that 120 men were still housed in a three-tier stack of 'portacabins', and the inspector observed, 'Should an emergency arrive, it would be very difficult to evacuate this installation quickly and safely. . . .' Similar misgivings were shared by one of his superiors, who baldly noted on the file's minute sheet, 'They are *drilling* with 330 people on board!'

Some three weeks before the certification requirement was due to take effect, the platform was visited again, and it was noted that the first well was about to be perforated, with 214 persons still on board, and that, indeed, *production* was now expected to begin before the end of the month. The inspector expressed his surprise at the amount of work still to be completed in view of the fact that production was imminent, and even described the platform as 'still very much a construction site'. While acknowledging that the standard of housekeeping was good, given all the problems of concurrent drilling and construction, he could not hide his own doubts as to 'whether it would be sensible (from other than an economic viewpoint) to add production operations with their commissioning problems and increased hazards while still at the tail-end of the construction phase'.

But the platform received its Certificate of Fitness some ten days after the prescribed date. As the Department later told the company, 'The Northern North Sea platforms with their lengthy periods of concurrent construction, drilling and production had not been contemplated when the regulations were drafted', so Certifying Authorities were advised 'to virtually ignore constrction quarters and activities and [to] give partial certificates to "drilling" '. This had subsequently been extended to include production. Thus another inspector, visiting the platform approximately one month after the Certificate had been issued, confirmed that production had indeed commenced, although completion was, in his opinion, still months ahead. 'The efficacy of the Certificate of Fitness may be gauged', he suggested, 'from the number of construction workers still on the platform' – no fewer that fifty-five out of a total complement of 199. In voicing his own misgivings on this occasion, he demonstrated that the wider issues discussed in earlier chapters do indeed at least sometimes filter down to the day-to-day level of enforcement:

> The economic plight of the country and the political significance of producing oil yesterday are understood and appreciated, but, in my opinion, from a purely safety aspect, 'start-up' has been three months too soon. I state this so that no one is in any doubt that corners are being cut and calculated risks are being taken to obtain objectives. I myself consider that the risks are valid, but I would be failing in my duty if my peers are not aware of the situation. The position is that construction work is going ahead simultaneously with well drilling, oil/gas separation and pumping activities.

The story does not, however, quite finish there. In March 1976, some six months after the inspector's foray into the political economy of offshore safety, he visited the platform once again. As a result of what he found on this occasion, he felt compelled to write to the company's safety officer explaining that, notwithstanding the company's possession of a Certificate of Fitness, 'many disturbing features are still evident.' Inevitably, perhaps, item one on his list was the fact that the installation was 'still in the construction phase although producing'. Later in the same year, the Department attempted to regularize its general position on the question of temporary accommodation by issuing a Guidance Note to the Certifying Authorities. As from May 1977, no reduction in standards *vis-à-vis* fire hazards or escape routes was to be accepted, though a somewhat lower standard of creature comforts could be, and the presence of 'construction-quality accommodation' was no longer to be ignored when drilling and production were in progress. In the event, the company in question did make one last attempt to secure further exemption, a request to this effect being received via the Certifying Authority in the spring of 1977. This time, however, the Department appears to have dissimulated successfully about its continued willingness to grant exemptions 'in certain rare instances', and although the company was allowed some concessions in connection with temporary accommodation on another platform in the same field, it was confirmed that the Certifying Authorities had been instructed to enforce the regulations from May of that year onwards.[25] What happened thereafter is not known, although the issue of temporary accommodation did crop up once more in the context of gas flaring. Again, too, the Department's dilemma became clear, even if it did get its way in the end. This time, the story can be told from the company's point of view:

> Well, one of the problems is that we come under pressure from both sides. For example, we were flaring gas on X platforms. The Department of Energy said, 'You must stop flaring gas.' We said, 'OK, we could stop it today by stopping the oil flow.' 'Well, no, we don't really want that, but we want you to put maximum effort into getting these NGL – this is the gas liquid fraction – plants commissioned and operating safely.' We said, 'OK.' So we put a massive effort into that, and that means people on the platform. . . . So we put a massive effort into commissioning NGL for the Department of Energy, and then they come round and say, 'Ah! You've got some temporary accommodation there – we don't like that. We think you had better have permanent accommodation.' We say, 'Certainly! To do that we first of all have to get engineers out there to do the design study. We then have to get the old stuff off and the new stuff on. We need people there to do it.' So what we find now is that we've got the four platforms, and attached to two of these we have accommodation rigs. . . . We are sympathetic to what the Department is saying, but we can't really do everything. . . .

This case history does not lie very easily alongside the assurances which had been given with regard to the use of powers of exemption envisaged by the Mineral Workings (Offshore Installations) Act in 1971, when the Under-Secretary of State had insisted that exemption from Construction and Survey Regulations was only contemplated where 'parts of the operations on a rig might be covered in other safety legislation'.[26] Even in its generality, the Health and Safety at Work Act did not apply to construction work on the installation in question at the operative time, while the Under-Secretary's other candidates – legislation pertaining to ships, hovercraft and mining – can scarcely have seemed terribly relevant. But on one score at least, the Department was correct. Although the suggestion that the northern North Sea platforms had not been contemplated when the relevant regulations were being drafted almost certainly owes more to infelicitous phraseology that to historical fact, there is no doubt that the extent to which construction work would become protracted was drastically underestimated by officials and companies alike. In itself, however, this does not explain why 'calculated risks' were preferred to delay, and it is hard to escape the conclusion that the inspector's own explanation in terms of economic priorities was correct. Nor was he alone in admitting that enforcement could not always display total unconcern for such priorities. One of the HSE inspectors openly acknowledged such influences upon his own activities pertaining to pipelaying:

> The trouble is when they are offshore, it's always – it's quite often too late to do very much about it really. . . . It may be something fairly major you have got to do, and you either stop the pipelaying altogether, in which case there is a whole load of pressures come in on you – you know, 'You are stopping a bright new future that's facing us all' – or you live with it until the end of the next season. . . .

One last context in which wider forces left their imprint upon enforcement practice in the form of institutionalized tolerance arose out of the specialized administrative structure which, conflicts from 1976 onwards notwithstanding, has dominiated the regulation of offshore safety. As we saw in chapter 5, the emergence and perpetuation of this 'vertical' structure encompassing the Department of Energy and its predecessors not only owed much to the broader issues of political economy at stake in North Sea development, but also, as is so often the case in such matters, involved certain disadvantages as well as benefits. Among the former, and of considerable importance for the argument here, is the fact that the degree of specialization entailed by such arrangements can lead agencies to develop their own idiosyncratic institutional culture and to their departure from the generally established procedures followed in other parts of the bureaucracy.[27] As far as law enforcement is concerned, this tendency can manifest itself in the routine utilization of different recipes for securing compliance, in a different approach to the imposition of criminal sanctions and, indeed, in a different attitude towards the role of law in the whole enforcement process. These characteristics, I want to suggest, did indeed come to permeate the enforcement practice of the Department of Energy.

This tendency is apparent in a number of different ways. Most pervasively, perhaps, it is evident in the Department's day-to-day documentation procedures, which, unlike those of the Factory Inspectorate, for example, do not routinely utilize legal categories as an organizing framework for the presentation of relevant information. While the inspection and accident reports prepared by the PED are technically detailed and do point out omissions, necessary improvements and so on, they do not cite statutory requirements or regulatory codes in any standardized manner. Nor do the notes left behind with installation managers routinely refer, in explicit fashion, to legal requirements as the bench-marks against which deficiencies are measured or remedial action enjoined. As the following extract from another report illustrates, inspectors may even sometimes invoke their formally draconian power to 'require the owner or manager or any person on board or near to an offshore installation to do or to refrain from doing any act as appears to the inspector to be necessary or expedient for the purpose of averting a casualty'[28] without specifying the legal basis for such action:

1. The following entry was made in the official logbook: 'Personnel are not to work on the drill floor or on the BOP [blow-out preventor] deck without written permission . . . until the following safety hazards have been removed:
    1. Drill Floor – Rathole sleeve tube to be safely secured.
    2. BOP Deck – Part of the BOP stack suspended from knotted wire rope – arrangement unsafe – lower or re-rig safely as soon as possible.
    3. BOP Deck – Conductor pipe stacked very dangerously – make safe and secure.'

One explanation for this lack of orientation is, of course, that, like North Sea operations themselves, the attempt to apply safety legislation in that setting is comparatively recent, and there has not therefore been time for bureaucratization along legal lines to take place. Added to this, the perpetuation of a separate administrative structure for offshore affairs meant that exposure to the procedures used in other sectors of safety law enforcement onshore was minimized, with the result that the bureaucratically entrenched significance of legal categories within organizations such as the HSE had little effect. Whatever the cause, however, the outcome seems to have been a pattern of enforcement which accorded relatively little salience to law as the specific foundation for official action.

A similar sequence of cause and effect can be discerned in other aspects of the Department of Energy's enforcement activities which have an even more direct bearing on the question of institutionalized tolerance. The PED, as its director conceded in 1979, 'has not in the past been very successful in preparing evidence for prosecutions', an unsurprising shortcoming which he put down to the fact that, unlike their HSE colleagues, PED inspectors have not been able to devote something like two years to learning. Not is there any doubt that, on occasion, this lack of experience and training has shown itself. In one case where proceedings were considered but not ultimately taken, for example, the head of Operations and Safety had to seek enlightenment from the Procurator Fiscal's office as to the meaning of 'summary conviction', while his instructions to one of the inspectors involved in the case revealed a marked unfamiliarity with what was involved in preparing a case for prosecution. Thus he was 'uncertain as to what is the best way to go about drumming up this sort of evidence, and whether Legal Branch ought to have documentary evidence in their hands as to what happened'. Reflecting a broader departmental lack of experience in such matters, he went on to predict 'One thing at least this attempt [to prosecute] will achieve will be some sort of procedural note from me so that this sort of evidence can be obtained hot off the press rather than one month later as we now stand.' His closing comment displayed an element of that understandable levity with which the uninitiated approach an unfamiliar task – 'Perry Mason appears to have gone out of business but perhaps this new character Rockford could assist you.'

From what has already been said, it is clear that the Department of Energy makes little secret of its lack of expertise in the matter of legal proceedings. Indeed, the comments from the director which were quoted above only echo the views expressed to the author by one senior inspector, who readily admitted that he and his colleagues lacked the background and experience of Factory Inspectors in such matters, with the result that prosecution was rendered particularly problematic. Needless to say, however, outside observers were somewhat more caustic on this score. According to one not otherwise unsympathetic HSE man, for example, 'the Department of Energy people . . . have no experience of enforcement whatsoever', while he did not think 'they would have the first idea about how to go about gathering evidence'. More generally, this and other features of the Department's performance with regard to the expressly legal side of its brief came in for heavy criticism from the dissenting members of the Burgoyne Committee:

> Scruting of investigation reports on serious or fatal accidents has tended to indicate absence of some necessary factual information; insufficient consideration of legal aspects necessary to identify contraventions of laws or regulations; absence of first-hand statements taken by PED inspectors; delay sometimes considerable in assembling papers which are not normally accompanied by a covering report on the lines of the Factory Inspectorate accident report; and insufficient consideration being given to enforcement procedures generally or legal proceedings in particular.[29]

Reluctance to prosecute is not, of course, the exclusive prerogative of the Department of Energy. As suggested at the beginning of this chapter, the general ethos of law enforcement pertaining to the safety of working conditions in Britain is not one that attaches very high priority to legal proceedings as an appropriate response to infringement. Conversely, it is true that PED do acknowledge that such a step can have a very salutary effect in certain circumstances, and that there is no indication of criminal sanctions being relegated to quite the symbolic role accorded to them by one Certifying Authority which blithely explained, 'Penalties are necessary as a public demonstration of the application of the Act.'[30] No less important, the Department has a familiar argument in its favour, inasmuch as a fine of £400, the maximum punishment for most offences on summary conviction under the relevant legislation, scarcely constitutes an economically realistic deterrent to an oil company or its subcontractors.[31]

Bearing all these caveats in mind, however, it remains the case that the distinctive culture which has emerged within the Department of Energy seems to place even less store than its main onshore counterparts by the use of the courts. In particular, it has shied away from endorsing the approach of the HSE, which, despite a low rate of prosecution in absolute terms,[32] does nonetheless see this strategy as an important back-up to the inspection and investigation process. Thus, once again, it was the PED's director who voiced the Division's collective feelings when, in the context of comments upon the views of the HSE, he explained that his Inspectorate 'would strongly deprecate the attitude sometimes displayed of looking for breaches of regulations rather than ways to prevent similar accidents'. Similarly, another senior official in the Department dissented strongly from what he saw as a tendency to measure success under the Factories Acts and the Health and Safety at Work Act in terms of prosecutions. Nor does the Department even seem to be very willing to share information with the Executive in connection with such prosecutions as have taken place. Asked for assistance in tracing cases which have occurred, a member of the latter body had to admit that they were experiencing difficulty in getting up-to-date information on this score, much less information from the past.

Against this background, it is not perhaps surprising that the number of prosecutions arising out of the infringement of offshore safety regulations has been very low. With the help of the relevant authorities, particularly the Crown Office and the Procurator Fiscal Service in Scotland, it has been possible to trace only thirteen cases that reached the courts up to the end of 1980. Involving twenty-three separate parties, including operating companies, subcontractors and, occasionally, employees, all have occurred since the beginning of 1977. In once instance, the HSE appears to have been the instigating agency, in pursuance of its duties connected with pipelaying, while another case, which was ultimately 'deserted' by the prosecuting authority (in Scotland, the Procurator Fiscal rather than the enforcement agency itself), was later described as having been keenly supported by the HSE but not by the Department of Energy. Worth noting also is the fact that the extension of the Health and Safety at Work Act to the Continental Shelf seems to have played a major role in prosecution activity, no fewer than six of the thirteen cases involving charges under that statute. Sometimes too, when the reasons for wishing to initiate proceedings were recoverable from the files, they proved revealing in themselves. Extraneous considerations, coupled with a more commonplace desire to use prosecution for exemplary purposes, are evident in the following extract:

> In a previous case . . . regarding the failure to report a fatality . . . we accepted your advice not to prosecute [i.e., advice given by Legal Branch]. This fatality has now, for other reasons, become a *cause célèbre* in some union circles and, in hindsight, it was a mistake not to prosecute. . . . [There is], in our view, sufficient justification to contemplate prosecution of X as a warning to the rest of the industry.

The desire to warn was also present in another instance, though the need to do so scarcely reflected favourably upon the Department's earlier attitude:

> the motivation of the Department in submitting the case is that they wish to be seen to have the ability to institute a prosecution if they think fit. Since the establishment of the Diving Inspectorate, the Department has proceeded by exhortation, to obtain compliance with, first of all, logical safety measures and later, compliance with the Diving Regulations. The Inspectors feel now that since there have been no prosecutions, the companies may adopt the attitude that there will be no prosecutions.

The incorporation of something almost approaching a notion of corporate *mens rea* into decision making, a phenomenon common in some other areas of safety legislation,[33] is apparent in a third extract, as is consideration of some of the issues mentioned earlier in this chapter:

> X have a particularly poor safety record (albeit improving). They

have little appreciation of their responsibilities under British legislation and their management, in my view, has not come to terms with
the problems of running a dangerous operation offshore. They are
still of the opinion that accidents will always happen, even though
this particular one was easily avoidable by spending a little more
time (25–30 mins.) and using correct procedure. They admit that it is
difficult recruiting competent, trained personnel, yet their efforts to
compensate by adequate supervision are patently inadequate. If [the
operating company] feel that they cannot improve the situation any
further, I think a prosecution would impress upon [them] the serious
nature of the problem and also their obligations when they are
working on the UK Continental Shelf.

The thirteen cases to reach the courts are discussed in the third
section of this chapter, and further details of how they fared in that
context will be provided at that point. Suffice it to say here that the
evidence on prosecution once again supports the view that tolerance of
violation has been institutionalized at a comparatively high level. The
enforcement of offshore safety regulations has thus far been
dominated by an administrative structure which, for whatever reason,
developed a distinctively low-profile approach to the application of
legal sanctions against offenders. Not usually pursued in the courts,
and therefore shielded from public scrutiny, the vast majority of
offshore safety violations have become conventional crimes *par
excellence*.

If prosecution is unlikely, what other strategies does the Department
adopt in order to secure compliance? Here, a great deal is made of the
inspectors' power to close down an installation, a sanction which they
and the industry claim to be entirely adequate. Superficially convincing
as this claim may be, however, it still leaves certain important questions unanswered. Would this power, for example, be sufficient to
overcome the political and economic pressures to maintain continuity
of production? Given the Department's wider concerns, is it likely that
such a sanction would be used or threatened with any regularity? 'Very
rarely' was the estimate of one official, who went on to explain that
when a direct confrontation between production and safety had arisen
in connection with a number of the older gas platforms, the issue had
been resolved by the granting of temporary exemptions and partial
certificates. Other officials merely offered isolated anecdotal accounts
from their own experience of the dramatic impact which a threat to
close down an installation could have. Sometimes, as when an inspector described how he had threatened to take such action upon

discovering that an installation lacked a single life-saving capsule with
a startable engine, such stories revealed just how high the other price of
British oil could have been.

Most of all, however, the claims made for the efficacy of this
sanction tend to divert attention away from what, in the light of the
British North Sea's record so far, is at least an equally important issue –
the question of what action is taken in connection with all those other
mundane deficiencies which could result in injury, but which are not
deemed to warrant contemplation of such draconian measures. Here
again, the Department seems to plough its own furrow, having little
time for the complex system of improvement and prohibition notices
provided by the Health and Safety at Work Act and extensively used by
several of the HSE's operational divisions as a coercive alternative
stopping short of prosecution. According to the dissenting members of
Burgoyne, no notice of either type has thus far been served by the
Department in its role as agent,[34] while once again, an HSE man traced
the lack of interest in such measures back to basic inexperience in the
enforcement field. Instead, it seems that the Department relies upon its
own rather ad hoc follow-up procedures and, where an issue gives rise
to particular concern, upon the same inspectorial power to require
action or desistance on the spot.

In the absence of access to an adequate sample of sequentially
ordered inspection reports, it is obviously impossible to give a definitive answer to the question of how successful or otherwise this
idiosyncratic approach may be. From the available evidence, however,
there are at least some grounds for scepticism, as for the suspicion that
this system operates in such a way as to institutionalize tolerance of
violation even further. On one installation, for example, an inspector
found a damaged crane to which access could be obtained only by
climbing up the pedestal and clambering across the boom to the cab,
an operation which involved 'obtaining hand-holds by inserting
fingers in bolt holes in such a manner that loss of balance would have
meant amputating a finger'. Some measure of his on-the-spot action to
avert the rather obvious risk of a casualty in this context can be gained
from the fact that, just over a week later, he was telephoned from the
Department's London headquarters and told that he 'should have
instructed the OIM (offshore installation manager) to cease using the
crane'! On another occasion, an HSE inspector who had joined an
offshore visit on a courtesy basis recounted, the absence of certain
fittings which had first been requested several years earlier was noted.
When the approach which the HSE man would have adopted to this

situation – an improvement notice with a four-week time limit for completion – was pointed out, the installation manager rang his shore base for immediate action. Similarly, one of the files which it was possible to examine in the course of this project revealed a remarkable sequence of what must surely qualify as institutionalized tolerance. As an inspector observed at one stage in this protracted affair, the lack of overwind trips on cranes had already been pointed out some seventeen months earlier, while slippery and unsafe deck surfaces (the accompanying safety officer actually fell during the visit) and the need for safety chains at open wells had been drawn to the attention of management over a year before. Only eight months had elapsed since the company was told about insecurely fastened flameproof electrical installations, damaged fire baffles in ceilings and deteriorating anti-slip paint on the deck. Not surprisingly, he thought, 'It might be a good idea for London to write to X direct on these points.' Whether or not this was done cannot be established. What can, however, is that some three months later another inspector described his general reaction to conditions on the installation as 'one of sorrow rather than of anger', while nearly a year after that, the Department was still asking for overwind trips to be fitted to the cranes and for attention to be given to one of the causes of slippery surfaces specifically pointed out in the course of the earlier inspections.

The Causes and Consequences of Confusion

If institutionalized tolerance became one distinctive feature of the offshore safety regime's operation, a degree of confusion amounting almost to endemic chaos was to become another. Permeating the regulatory system in a number of different ways, this inevitably compounded the already acute problems of enforcement, attenuated the authorities' capacity to respond effectively still further and created a series of anomalies and loopholes of which, as we shall see, the industry was not slow to avail itself. Omissions, unclear definitions, disarticulation between different segments of the bureaucracy and even dubious jurisdiction on some issues were all to play their full part in rendering the effective application of UK legislation to the Continental Shelf doubly problematic in practice. While limitations of space preclude detailed analysis of all the difficulties which arose in this context, no account of how offshore safety regulations have been enforced would be complete without some discussion of the major factors involved.

That some problems of this kind should have arisen was, of course, only to be expected. After all, new bodies of law characteristically experience teething troubles, and the development of foolproof enforcement procedures is typically, at least for a time, a matter of trial and error. Thus, to take an example cited above, it is neither particularly astonishing nor significant that the framers of the Health and Safety at Work Act 1974 (Application outside Great Britain) Order of 1977 should have overlooked the fact that the legislation now being extended to the Continental Shelf provided no powers for requiring transportation to offshore installations.[35] While the immediate consequences may indeed have been dire – in the words of one senior police officer, HSE inspectors were left 'more or less high and dry' – it is with such ommissions that enforcement agencies all too often have to contend, particularly in the aftermath of new legislation. Similarly, failure to cover all possible contingencies is an ineluctable feature of the legislative process, as is the fudging of definitions which subsequently turn out to be pivotal in the law's operation. In the latter context, for example, the meaning of the phrase 'capable of being manned', crucial to the implementation of the Health and Safety at Work Act *vis-à-vis* the interface between pipelines and installations and to some parts of the Mineral Workings (Offshore Installations) Act of 1971, was left very unclear and was to become a bone of considerable contention.[36] Indeed, just as a well-known sociological treatise on horse racing once devoted an entire chapter to the question 'What is a horse?' so a work orientated more than the present one to the legal solution of legal problems could quite justifiably allocate a substantial section to asking 'What is an offshore installation?'[37]

At a later point we shall see that companies involved in North Sea operations did not hesitate to invoke the existence of such ambiguities when the occasion necessitated. But on top of these understandable side-effects of the normal law-making process, there were also other problems emanting from a more substantial basis than inadvertence or the lack of legislative prescience. At one level, jurisdictional issues stemming from Britain's ambivalent stance with regard to international law on the subject of the Continental Shelf, not just legal niceties but questions stretching right back to the international dimension of political economy in a way that cannot be fully expounded here, spawned important practical problems which have not yet been fully resolved. Closer to home, because the Continental Shelf Act of 1964 was passed in an aura of unconscionable haste aimed at the early facilitation of a viable system of proprietary rights in the mineral resources of the Continental Shelf, it signally failed to make adequate provision for the extension of some parts of the law which, although of vital relevance to the issue of safety, did not fall conveniently inside the formula of criminal and civil law utilized in the drafting of Section 3.[38] Not least, and despite some vigorous protests, it took much less than adequate cognizance of the fact that future offshore activities could conceivably fall within the jurisdiction of Scots law. As in legislation, so too in practice the coherence of the regulatory system was to suffer badly from an arrangement whereby, while control was organized from the centre, its implementation would often depend, at the end of the day, upon the legal system prevailing in what appears to have been treated as the periphery.

To understand the nature of the problems emanating from the international plane, it is necessary to go back not to the Continental Shelf Act but to the Convention on the Continental Shelf which preceded it.[39] By Articles 5.2, 5.3 and 5.4 of this international agreement, coastal states were empowered, among other things, to establish safety zones of up to 500 metres around installations, to take necessary measures within those zones for the protection of such installations and to exercise what has been interpreted as full jurisdiction on the installations themselves. In some respects at least, the UK has been scrupulous in abiding by the terms of this agreement. Thus, for example, the powers of police constables as defined by Section 11(3) of the Continental Shelf Act are restricted to installations,[40] and although Section 2 of the same Act does not specify the geographical extent of the safety zones which may be established under its authority, in practice this has been restricted to 500 metres.[41] Moreover, although the cumbersome nature of the procedures involved may be largely responsible, the UK has not put the Convention's ambiguity on the status of mobile drilling rigs to the test by establishing safety zones around these installations.[42]

In other respects, however, the propriety of the UK's actions in international law is more questionable. As Grant points out in his annotation of the Convention, the extension of UK criminal law to an area of 500 metres around every installation under Section 3(1) of the Continental Shelf Act is of dubious legitimacy on several different scores.[43] For one thing, the inclusion of installations which have not been made the subject of safety zones under Section 2 of the Act is problematic, while the powers conferred by the Convention in relation to such zones are limited to those necessary for the protection of the installation. The restriction of general jurisdiction to the actual installation itself under Article 5.4 of the Convention raises further difficulties, as does Article 2.1, which confers only such sovereign rights over the Continental Shelf as are necessary for the purposes of exploration and exploitation. Similar problems arise in connection with regulations made under the Mineral Workings (Offshore Installations) Act with regard to the implementation of Section 84(3) of the Health and Safety at Work Act, the provision allowing for the Act's extension outside Great Britain. As the same author observes, these may contravene the Convention if they purport to extend control beyond installations or beyond the regulation of safety-zone activities which are undertaken by British citizens, bodies corporate and ships or which contravene measures for the protection of the zones. The application of the Health and Safety at Work Act to pipeline works would seem to be a case in point.[44]

From such stuff are the minutiae of test problems in international law fashioned. In the present context, however, the point is that these problems also created some serious obstacles to the development of an effective and comprehensive regime for the control of offshore safety. Thus, for example, while there are cogent practical arguments for enlarging safety zones to protect the bigger installations which have been necessitated by more northerly developments, not to mention semi-submersibles which may have anchors located well beyond the 500-metre limit, international law has thus far presented an insurmountable obstacle to such a step.[45] Nearer to the enforcement issue itself, grave jurisdictional problems arise in connection with the powers and protection afforded to the police, who not only would have to provide the ultimate coercive back-up to the safety regime in extreme situations, but also have an independent role to play in connection with criminal law and the investigation of fatal accidents. Whatever jurisdictional doubts there may be about their powers within the 500-metre zone, and it is by no means certain that such do not exist, outwith that area their position is almost impossibly anomalous. Thus, while the Health and Safety at Work Act empowers inspectors to take a constable with them if there is reasonable cause to anticipate serious obstruction,[46] it is extremely doubtful whether the police would be justified in complying with such a request if such an eventuality should occur in connection with, for example, a pipelaying barge operating outwith the 500-metre zone. Moreover, as is apparent from the following summary of an incident recounted to the Burgoyne Committee by the Chief Constable of Grampian Police, jurisdictional problems associated with the enforcement of criminal law could have potentially disastrous implications for safety.

In 1976, Grampian Police were informed that five alleged incidents of fire-raising had taken place during one night on a barge which was Panamanian-registered, owned by a Dutch company, on charter to an American oil company and operating within the safety zone of a production platform in the British sector of the North Sea. At the time of the incidents, however, the barge had been lying outside the safety zone because of adverse weather conditions, and the question immediately arose of whether the police had the requisite powers to respond to the master's understandable call for assistance. After consultations, it was agreed that because the barge was Panamanian-registered and had been in international waters at the operative time, Grampian Police had no jurisdiction. The Dutch authorities also concluded that they had no power to intervene. However, it was agreed by all concerned (though not presumably by the fire-raisers) that the police had a 'moral obligation to attend', and, after an investigation involving five officers for two days, the matter was resolved, two crew members being taken off and 'returned' to Holland. As the Chief Constable observed, although the moral obligation had been accepted, it appeared that the police had been acting *ultra vires*. Nor could he be confident that further and possibly even more serious incidents of this kind would not occur in the future. It did not seem terribly logical, he concluded, that the barge was subject to UK criminal law while within the zone but to Panamanian law when outside, any more than it was consistent that some statutory regulations (for example, the Health and Safety at Work Act) should purportedly apply to such vessels in the latter circumstances, while 'the criminal law and police powers are not enforceable.'[47]

The eventual solution to problems of this kind would lie, of course, in the revision of relevant international agreements. At this level, however, we encounter the broader analogue of the issues of political economy which were discussed in chapter 4. The United Nations Conference on the Law of the Sea has staggered along for some seven years, not just because the legal issues are extremely complex but also because it has become something of a test case with regard to the extent to which the Third World will become a full participant alongside the major powers in the global negotiation of a 'new international economic order'. The United States is currently said to be on the brink of withdrawing from her previous stance, even though final agreement was thought to be imminent, while according to one British authority on such matters, the Conference has anyway become more obsessed with issues like the new bonanza to be expected in deep-sea mining than with the residual jurisdictional chaos surrounding oil operations in areas such as the North Sea.[48] Not surprisingly, too, the conflicts of interest which permeate negotiations on the international plane filter down and lend an added dimension to the internal politics of the state discussed above.[49] Enigmatically, the Chief Constable of Grampian Police saw one of the main obstacles to effecting necessary amendments to the Continental Shelf Act as 'the reaction of certain government sources' which regard any extension of British criminal law 'as interfering with the right of passage of vessels on the high seas'.[50] With reference to the enlargement of safety zones, the Department of Energy was less coy in explaining the predicament to the Burgoyne Committee:

> The Secretary said that [the Department of Energy] recognized the main problems highlighted by UK Offshore Operators' Association in relation to [safety zones] . . . . However, strict interpretation of the international convention and the difficulty of persuading other departments – Foreign Office and Department of Trade – of the need to change that part of the convention (interpreted by them and others as an increase of coastal state powers over areas of high seas) had so far frustrated efforts for larger or more sensible configurations of zones.

While the problems discussed thus far were certainly serious enough, they became comparatively minor when set alongside the difficulties which were to emerge with regard to the investigation of fatalities occurring in the North Sea (or, more precisely, in that part of it which falls within Scottish jurisdiction) by authorities other than the inspectors themselves. Here one of the principal causes of confusion was that whereas the jurisdiction of an English coroner in connection with the holding of inquests is founded upon the place where the dead body is lying or is first brought to land,[51] for the greater part of the period during which offshore operations have been taking place the Scottish Sheriff's power to hold a Fatal Accident Inquiry – and the powers of Procurators Fiscal and police to carry out the investigations prerequisite to such an inquiry – did not rest upon such a convenient formula. Founded upon separate legislation for Scotland, these powers therefore depended upon whether the relevant statutes could be deemed to apply to the particular circumstances and location of North Sea fatalities. As a result, the issue became a fruitful hunting ground for those skilled in manipulating the omissions and ambiguities which the Continental Shelf Act and associated legislation had bequeathed. Indeed, for a time, such efforts were so successful that the jurisdiction of the Scottish courts in this matter was virtually placed in suspension.

Under the Fatal Accidents Inquiry (Scotland) Act of 1895 and the Fatal Accidents and Sudden Deaths Inquiry (Scotland) Act of 1906, deaths resulting from an accident during industrial employment were made subject to compulsory inquiry, while the Lord Advocate could, at his discretion, order such an inquiry into any sudden or suspicious death where he considered this course of action to be in the public interest.[52] Section 3 of the Continental Shelf Act, it will be recalled, purported to extend the criminal and civil law of the UK to acts and omissions taking place on or around installations in designated areas.[53] The Continental Shelf (Jurisdiction) Order 1968 both defined the Scottish border in connection with the areas thus far designated and, by Articles 2 and 3, conferred jurisdiction on the law and courts of Scotland with regard to acts and omissions occurring in the Scottish area thus defined.[54] All this done, it might be assumed that a watertight system had been constructed. Such, however, was to prove far from being the case.

The first intimation of what was to come seems to have been given in 1973, when a diver was killed in the course of a dive being undertaken from a pipelaying barge. On this occasion, the Procurator Fiscal's power to investigate the accident was challenged – by telephone – on the grounds that Section 12(3) of the Mineral Workings (Offshore Installations) Act expressly excludes from the definition of an offshore installation one 'which at the relevant time consists exclusively of a pipeline, whether or not any part of it previously formed part of an offshore installation'. Hence, so the argument ran, any attempt to claim jurisdiction by dint of the Continental Shelf Act's references to offshore installations in designated areas would be invalid. Undaunted, however, the Fiscal stuck to the view that insofar as the provision of Section 12(3) did exclude pipelines, it did so only for purposes of the 1971 Act itself, the general jurisdiction conferred by the Continental Shelf Act and attendant Orders remaining unimpaired. The investigation accordingly went ahead, although it was not helped by the fact that early doubts as to jurisdiction had led to failure to impound the diver's helmet. The company, obviously anxious to establish for itself what had happened, returned this piece of equipment to the parent company in the United States for investigation. Unhappy coincidences of helicopter unavailability, followed by an unexpectedly fast turn-around when the barge next touched port, created further investigative problems. At the time when evidence relating to this part of the project was being collected, there was no record of a Fatal Accident Inquiry having been held.[55]

Jurisdictional problems recurred in the following year, 1974, though this time an inquiry was held. Again, a diving accident was involved, a New Zealander having been lost after a dive to attach a pennant line to an oil rig's anchor cable. One complication was that his body had not been recovered, but the major source of doubt was the question of whether the rig, which had been under tow to its drilling location at the time, could be classed as an installation. The latter term, the Procurator Fiscal pointed out to the Crown Office (the body comprising the Scottish Lord Advocate's permanent staff under the command of the Crown Agent and responsible for direction of the Fiscal service) had not been defined in the Continental Shelf Act. Furthermore, although the Mineral Workings (Offshore Installations) Act could certainly be interpreted as including 'floating installations' in the relevant definition,[56] some of the Regulations made thereunder seemed to be more ambiguous, inasmuch as they exempted 'installations registered as vessels . . . which are in transit to or from a station'.[57] The rig in question was registered as a vessel with the US Coastguard. Accordingly, he asked Crown Office to secure instructions from Crown Counsel as to the competency of holding a Fatal Accident Inquiry under the Acts of 1895 and 1906. Whether or not his doubts were shared in Edinburgh it is not possible to say, but the instructions which he subsequently received were clear enough, and the inquiry went ahead.

Although no challenge was offered on this particular occasion, the jurisdictional gauntlet was thrown down in no uncertain manner shortly thereafter. In August 1974, a Spanish welder died when an anchor buoy on a pipelaying barge crushed him against a rail. The barge, which was owned by one of the largest offshore contracting companies, was engaged in laying a pipe between a well head on the sea bed and the coast. Possibly because of his earlier encounter with the jurisdictional vagaries surrounding pipelaying barges, the Procurator Fiscal this time conceded freely that such vessels did not come within the definition of an offshore installation. But his ingenuity was not thereby exhausted. Arguing that during the construction phase a pipeline is manned and is part of, or at least connected to, the well head, he claimed jurisdiction on the grounds that in these circumstances it qualifies as 'an offshore installation to which the Acts apply'. Unfortunately, however, his ingenuity was to founder on the rocks of unassailable fact – at the operative time bad weather had caused the pipeline to be disconnected, the investigating police officers being understandably enough misled by the further fact that it had been reconnected by the time of their inquiries. The company's legal department made the point with fatal precision:

> Even assuming that the right exists under the Fatal Accidents Inquiry (Scotland) Act 1895, which is not admitted, the writer understands that you claim the right to hold the inquiry on the ground that the death occurred within 500 metres of an offshore installation. For this purpose it is your contention that the pipeline was an offshore installation within the definition contained in the Mineral Workings (Offshore Installations) Act 1971 and that the provisions of Section 12(3) thereof do not relate as the pipeline was not at the relevant time *exclusively* a pipeline, being a pipeline in construction attached to the pipelaying barge. This argument is, however, factually incorrect as at the relevant time, and, indeed for some days prior thereto, the pipeline had been laid on the sea bed and was in no way attached to the barge. Accordingly, insofar as a pipeline in the course of construction is an installation, and, again, this is not admitted, at the relevant time it was exclusively a pipeline and, therefore, not an offshore installation, and the right to hold a public inquiry claimed on the ground of an accident occurring within 500 metres of an offshore installation is without foundation.

Worth noting here is the fact that the legal process being objected to is one that takes the form of judicial inquiry rather than a criminal trial, one that normally takes place only after questions of criminal liability have been dealt with at such a trial, if appropriate, and one that protects witnesses from the jeopardizing consequences of incriminating evidence.[58] In effect, the objection is to what, despite procedural differences and some definite advantages, is the Scottish equivalent of an inquest. Though effective enough, moreover, the challenge on the above grounds was not the only string to the company's jurisdictional bow. Quoting advice received earlier through the Scottish Courts Administration, the legal department was able to raise a much more devastating objection:

> 'It does not appear to be clear that the Fatal Accidents Inquiry (Scotland) Act 1895 authorizes the holding by the Sheriff of an inquiry into the death of a person on an offshore installation situated in the Scottish area, as defined in the Continental Shelf (Jurisdiction) Order 1968. Inquiries under the 1895 Act are, strictly speaking, neither criminal nor civil proceedings and it may be that Section 3 of the Continental Shelf Act 1964, as amended, does not confer powers

> wide enough to enable the 1895 Act to be applied by order to these installations. In any event, the 1968 Order does not contain provision to that effect.'
>
> In the light of this statement we shall be obliged if you will kindly indicate on what ground the right to hold the public inquiry is claimed.[59]

Faced with these objections, the authorities decided not to proceed with the actual inquiry on the day which had been fixed, but to try for a debate on the jurisdictional question before the Sheriff on the same date. Crown Counsel's view was, 'If a decision is to go against us on jurisdiction, then the sooner this is decided the better.' The company, however, wanted informal discussions first, and on the appointed day no appearance was made for any party. Since the Sheriff was unwilling to adjourn the diet other than to a fixed date, the Procurator Fiscal concluded that this would only complicate the situation further, and the case was therefore not called. Apart from the obvious consequences in terms of the fatality in question, it was now clear that the jurisdiction of the Scottish courts *vis-à-vis* offshore fatalities in general was in almost complete disarray.

As a result of this case, all Fatal Accident Inquiries concerning deaths occurring on the Continental Shelf were suspended.[60] An attempt to remedy the situation was made by the Continental Shelf (Jurisdiction) (Amendment) Order 1975, but this order was not without its own deficiencies, and the powers which it conferred were not invoked.[61] Instead, a special section dealing with the Continental Shelf was built into the Fatal Accidents and Sudden Deaths Inquiry (Scotland) Act 1976:[62]

> 9.   For the purposes of this Act a death or any accident from which death has resulted which has occurred –
>   (a) in connection with the exploration of the sea bed or subsoil or the exploitation of their natural resources; and
>   (b) in that area, or any part of that area, in respect of which it is provided by Order in Council under Section 3(2) of the Continental Shelf Act 1964 that questions arising out of acts or ommissions taking place therein shall be determined in accordance with the law in force in Scotland, shall be taken to have occurred in Scotland.

This enactment allowed Fatal Accidents Inquiries to be resumed. Moreover, with its additional provision that the Lord Advocate could order an inquiry into deaths which had occurred up to three years before the Act took effect,[63] the way was opened to making up the ground lost during the intervening period. Thus, after what inevitably turned out to be a substantial gap in many instances, the backlog of suspended cases was slowly made up. In six cases, the time which had elapsed between the death and the holding of an inquiry was more than two years, while as late as mid-1978, six further fatalities which had taken place before the 1976 Act came into force still had not gone through the process. From investigation carried out at the former point in time, it seemed that most, if not all, of them were unlikely to do so.[64]

Although the 1976 Act removed some of the most obvious jurisdictional flaws affecting the investigation of offshore fatalities by the Scottish authorities – no longer, as under the stopgap Order of 1975, for example, is the Sheriff's jurisdiction restricted to deaths which are attributable to acts or ommissions – it by no means put a complete end to the potential for confusion in this crucial area of the safety regime. Would Section 1(1)(a), which makes investigation and inquiry obligatory where an employee has died as the result of an accident in the course of employment, cover an off-duty oil worker who is obviously obliged to spend the time between work periods on board the installation or an attendant accommodation rig?[65] More serious still, while the words 'in connection with the exploration ... or the exploitation of their natural resources' (Section 9) might seem to cover all contingencies, would it include operations connected primarily with the supply and maintenance of installations? Nor is this question merely idle speculation as to what residual gaps may still remain in the law. The issue became quite concrete shortly after the Act was passed, when an unsuccessful attempt was made to argue that there was no jurisdiction in connection with the death of a boilermaker who drowned while being ferried by Zodiac from a moored oil tanker to an installation for purposes of subsequent helicopter transportation back to land. The tanker had been undergoing boiler repairs, and it was argued that since the deceased had been employed on this task, he had not died as the result of an accident connected with exploration or exploitation.

Most of all, however, those involved still seem to be confused about the extent to which jurisdiction may yet be hostage to the unhappy terminology of the Continental Shelf Act and of the Orders in Council made thereunder. While Crown Office is said to take the optimistic view that Section 9(b) of the 1976 Act gives jurisdiction anywhere within a designated area to which Scots law has been applied by Order in Council under Section 3(2) of the Continental Shelf Act, and so indeed it would seem to say, others are less certain. According to one

Procurator Fiscal, for example, it is still questionable whether he has the requisite powers outwith the 500-metre zone specified in Section 3(1) of the same enactment, all the old problems of pipelaying barges and so on thus remaining unresolved. For this view, furthermore, he might be able to draw some pretty cogent support from the wording of that part of the relevant Order in Council which deals with the application of English, Scottish and Northern Irish law to designated areas:

> 2. *Subject to Section 3(1) of the Act*, the law in force in England shall apply for the determination of questions to which section 3(2) of the Act refers arising out of acts or omissions taking place in the English area; the law in force in Scotland shall apply for the determination of such questions arising out of acts or omissions taking place in the Scottish area. . . .[66] (Emphasis added)

As suggested above, one major factor behind the jurisdictional chaos which so nearly overwhelmed the Scottish authorities' power to investigate offshore fatalities was the Continental Shelf Act's failure to take adequate cognizance of its implications for the separate legal system of Scotland. However, it was not just at the legislative level that this lack of articulation between centre and periphery created confusion in the operation of the offshore safety regime. Additionally, the Department of Energy's inspectors, who had, of course, been able to pursue their own investigations into fatalities and other matters under the Mineral Workings (Offshore Installations) Act, had to familiarize themselves with the requirements of a different legal tradition, while their London headquarters had to come to terms with the fact that in Scotland bodies such as the Department of Energy and the Factory Inspectorate are rather less their own masters in matters legal than they are in England. The decision to prosecute, for example, rests with the Procurator Fiscal,[67] although the request for such action normally has to come from the relevant enforcement agency, and he also acts as prosecutor in the Sheriff court. On top of that, as we have seen, he has the responsibility for investigating fatal accidents, for reporting on cases to Crown Office and for calling an inquiry before the Sheriff, if this is deemed necessary. In the investigative context, the police act as his agents and therefore conduct inquiries in parallel (usually in collaboration) with any specialized regulatory agency that may be involved.[68]

Needless to say, this system creates a number of procedural interfaces, where there is room aplenty for confusion if a centralized agency is either not versed in its workings or less than enthusiastic about co-operation. Such difficulties surfaced in connection with the offshore safety regime within a year of the Department of Energy's coming into existence. Following the death of a roustabout at the end of 1974, solicitors acting for the dead man's father complained to the Lord Chancellor in England because the case had still not been resolved some nine months later. Appropriately, this complaint was passed to Scotland's Lord Advocate, and Crown Office immediately requested an explanation from the Procurator Fiscal in question. His reply revealed that there was a serious breakdown in communication between himself and the Department of Energy, arising mainly from the latter's unwillingness to supply copies of written reports. However, steps were being taken to improve the situation, and a few weeks later he was able to report on a meeting that had taken place with the appropriate official at the Department:

> I put it to Mr X that I have been impeded in progress of investigation of cases by not having access to the results of the technical investigations made by his Inspectors, and by my having been unable to receive written reports indicating whether or not proceedings were likely to be contemplated and which were required to be instituted through the Department.

What was at issue here was the Fiscal's role in connection with fatal accidents and prosecution. In the former context, he would normally expect to have the benefit of reports prepared by the regulatory agency, obviously more expert on the technical front than those prepared by the police, while he could decide on the question of prosecution only after receiving reports from the Department as to whether such action was desired. Most confusing of all, since a Fatal Accident Inquiry would usually be held *after* any criminal proceedings had taken place (the order is the reverse of that normally followed in England), he could not readily proceed with such an inquiry until the information requisite to the prosecution issue had been received and acted upon. Thus arose a procedural 'Catch-22', which was still in evidence late in 1976, when the head of PED was interviewed as part of a preliminary project to assess the feasibility of this research. London was waiting for the results of 'inquests' from Scotland before deciding whether or not to ask for the institution of legal proceedings; Scotland was waiting to hear about the possibility of such proceedings before going ahead with Fatal Accident Inquiries. While the discussion referred to earlier may have done something to clear the air, they did not put an end to the confusion. In April 1977, the Fiscal was complaining again:

> I have received a letter from Crown Office informing me that the Department have been in touch with them about this case and it seems to me that there may be some confusion as to the procedure being followed. As you know, I have not yet received a report on possible contraventions of regulations in this case which were mentioned to me by X [an inspector] some weeks ago when he visited me. I understand from you that a report was to be transmitted to me after reference to your Legal Department and that it would come to me either direct or through Crown Office. It is normal practice to dispose of any prosecution before the Fatal Accident Inquiry is held and consequently I expect to have both the police investigation and the report on the statutory matters coming to your notice at the same time, before decisions are made as to proceedings and Inquiry. I shall be obliged if you will look into the matter and let me know if there is some difficulty of procedure holding up the forwarding of the report.

While the Scottish system pertaining to the procedural sequence to be followed after fatal accidents certainly became the locus of the most acute confusion on this level, it was not, however, the only source of such difficulty. At one early point, for example, a PED inspector explained in interview that prosecution would often be impossible, even if it were desirable, because of the requirements of corroboration under Scots law, requirements which he wrongly interpreted as meaning that it would always be necessary for *two* inspectors to be present at the time of the offence's detection.[69] Similarly, difficulties of establishing the identity of the concession owner in one of the cases mentioned above were compounded by the Department of Energy's refusal to comply with a Fiscal's request for a copy of the lease in question. Even in the context of fatalities themselves, always the most likely kind of case to stimulate consideration of legal proceedings, there was for a time confusion over whose responsibility it was to lay the groundwork for possible prosecution. 'This we have left to the police, because the police force investigate and the coroner's inquest decides whether there are grounds for the prosecution,' explained the head of PED, echoing the procedural confusion outlined above. 'But the police don't go out there looking for offences,' said a Procurator Fiscal, 'When the inspector goes out, he is looking for offences.' More frequently, however, the complaint was that decisions and advice about prosecution which should have been solicited from

Scotland were being appropriated by London. A Crown Office official summed up the problem in connection with one case:

> If the Department of Energy proposed a prosecution, they don't appear to have written to us about it. . . . It's the Legal Branch of the Department of Energy that have given him advice, not us. It must never have reached us. They never know the Scottish procedure. If they want proper advice on whether a prosecution should take place in Scotland, then they should ask us and not some branch of their own Department.

Chaos in Court

Against the background of the issues covered in the previous two sections, it is appropriate to conclude this chapter with some comments on those cases which did run the gauntlet of the enforcement process from detection to court proceedings, however brief. Such proceedings, it will be recalled, have been comparatively rare, only thirteen traceable cases having followed this route from 1977 to 1980 – three in the former year and in the year following, two in 1979 and five in 1980. While the discussion here, therefore, can only be fairly brief, the record of these cases nonetheless underlines once again the peculiar difficulties which have attended the enforcement of offshore safety legislation. Not least, as can be seen from table 10, it shows that even when the authorities are sufficiently confident to embark upon this course of action, their success is by no means guaranteed. No fewer than five of the thirteen cases which reached the courts resulted in no finding of guilt against any party, while two of the three defendants in a further case were found not guilty. Of the twenty-three companies or individuals involved, only ten were found guilty, nine of them being fined and one – the defendant in a test case – being admonished.[70] Only two of the findings of guilt followed a not guilty plea.

Some of the most crucial problems involved in seeing matters through to succesful prosecution became apparent in the very first of the above series. This case arose out of an accident in which two divers died during a shallow swimming dive to recover a drilling rig's anchor pennant, which had come adrift during transit to the drilling location. A decision to prosecute both the company and the diving supervisor was taken, several charges being preferred against each. Right away, however, difficulties arose in connection with pre-trial procedures.

TABLE 10
COURT PROCEEDINGS UNDER OFFSHORE
SAFETY LEGISLATION, 1977–80

Case no.	Classification of company or individual involved	Statute or regulations under which solely or principally charged	Outcome
1	Diving company	Offshore Installations (Diving Operations) Regs. 1974	£200 fine
	Diving supervisor	„ „ „ „	£25 fine
2	Diving company	Submarine Pipelines (Diving Operations) Reg. 1976	Deserted
3	Exploration company	Offshore Installations (Registration) Regs. 1972	Admonished
4	Diving company	Offshore Installations (Diving Operations) Regs. 1974	£250 fine
5	Pipelaying company	Health and Safety at Work Act 1974	£400 fine[a]
6	Operating company	Offshore Installations (Occupational Safety, Health and Welfare) Regs. 1976	£200 fine
7	Scaffolding company	Health and Safety at Work Act 1974	£400 fine
8	Diving company	Offshore Installations (Diving Operations) 1974	£300 fine
	Diving supervisor	„ „ „ „	£50
9	Operating company	Health and Safety at Work Act 1974	Dismissed as irrelevant
	Scaffolding company	„ „ „ „	Not guilty
	Crane company	„ „ „ „	£100 fine
10	Operating company	Offshore Installations (Occupational Safety, Health and Welfare) Regs. 1976	Deserted
	Scaffolding company	Health and Safety at Work Act 1974	Deserted
11	Casing crew suppliers	Health and Safety at Work Act 1974	Not guilty
	Drilling contractors	„ „ „ „	Not guilty
	Employee of casing company	„ „ „ „	Deserted
12	Casing crew suppliers	„ „ „ „	Dismissed as incompetent
	Operating company	„ „ „ „	„ „
	Drilling supervisor	„ „ „ „	„ „
	Drilling contractors	„ „ „ „	„ „
13	Suppliers of diving services	Submarine Pipelines (Diving Operations) Regs. 1976	Not guilty by Sheriff's direction to jury[b]

[a]The amount of the fine in this case is as reported by HSE. The file was not available for checking.
[b]This was the one case taken on solemn as opposed to summary procedure.

Although the company, which was foreign, had a UK associate with offices on the British mainland, only shore-based personnel were employed from there, and a prosecution launched in that direction would almost certainly have run foul of a defence to which we shall have cause to return at several points – that they were not the employers of the divers. Reflecting the hopelessly tangled web of employment and contractual arrangements which have ensnared the British authorities on more than one occasion, the situation was further complicated by the fact that among the papers handed over by the company was a form bearing the name of yet another company and purporting to be an application for employment from one of the deceased. Of practical import too was the question of where to serve the complaint, since it was statutorily necessary, as far as the company was concerned, to do this at a place of business of the accused. In the end, the Procurator Fiscal had no alternative but to request the police to dispatch an officer to hand the complaint to the senior employee of the foreign company on the rig itself. That the day-to-day operational distinction between the foreign company and its UK associate was rather less than the separation maintained in law can perhaps be inferred from the fact that the senior employee in question initially invited the constable to take the complaint back and to hand it in at the UK office. Pre-warned, however, the officer pointed out that this course of action would only be greeted by the response that no one from the relevant company was at those premises, and the complaint was duly served on the rig. The case proceeded to court, and, the company and the supervisor having each pleaded guilty to one charge, the others were dropped. The fines of £250 and £25 which were imposed underline the ludicrously inadequate level of the penalties which are imposed in this area of safety legislation, as in others.

Even less satisfactory was the outcome of a prosecution initiated against another diving company in the same year (Case 2). Again, the case arose out of a fatal accident involving a diver. Among the alleged offences were failure to appoint in writing a competent person to supervise the dive, failure to meet the requirements for the provision of stand-by divers, failure to provide plant and equipment necessary for the safe conduct of the dive, failure to ensure that the diver was secured by a lifeline or breast rope and failure to provide him with adequate means of communication. When the case to adjudicate upon this substantial catalogue of alleged neglect was called in court, it was continued without plea on the motion of the agent for the defence. The company subsequently made representations that it was not in fact the employer of the dead man, and when the diet was resumed in court, the case was deserted *pro loco et tempore*.[71] Of some importance to what follows at a later point in this discussion is the suggestion that, at this stage, the Procurator Fiscal alerted the Department of Energy to defects in the Submarine Pipelines (Diving Operations) Regulations 1976 with regard to the question of divers' employers.[72]

Problems pertaining to residual lack of clarity in statutory definitions lay behind the next case, which was embarked upon in order to establish whether, under the Offshore Installations (Registration) Regulations 1972, there was an obligation to register an installation while its jacket was still under construction. The company was duly admonished, and the issue was hopefully put beyond question for good.[73] In Case 4, however, the labyrinthine complexities of the question of who employs whom in the North Sea cropped up once more. Involving one of the diving companies against which proceedings had already been initiated on a previous occasion, this prosecution stemmed from a non-fatal incident in which a diver had received an electric shock and burns when the insulation of the heating element in his suit deteriorated, and the element broke and short-circuited. Alerted to the risks by the abortive case referred to above, the Procurator Fiscal instructed the police on this occasion to make inquiries of the company as to who had been the actual employer, and the ensuing case resulted in conviction and a fine. Interestingly enough, in this instance the Department of Energy had pointed out to the Fiscal that the Health and Safety at Work Act now applied, and that he might consider prosecuting under its provisions, although the Department had not consulted the HSE, the 'enforcing authority' on the subject. Clearly, the more rational administrative arrangements announced in 1976 and described in chapter 6 had not produced particularly close collaboration on the question of prosecutions, despite the HSE's substantially greater experience in this field.

Case 5 saw the direct involvement of the latter body as a result of its own responsibilities in connection with pipelaying. A man had been lost after being swept overboard from a barge, and, until the holding of a Fatal Accident Inquiry, the accident seems to have been attributed to yet another freak wave. At that point, however, evidence emerged to the effect that a guardrail had been missing, and, at the instigation of the HSE, proceedings were taken under Section 2(1) of the Health and Safety at Work Act. This was the one instance encountered in the series when, despite the normal Scottish practice of holding criminal proceedings first, the preservation of the Crown's right to prosecute after such an inquiry was utilized.[74] Although no such departure from the usual procedure in this respect was necessitated by the following case, it did nonetheless throw up some other problems, one of which was subsequently to prove fatal to the prosecution in another case. An improperly secured grating had given way on a production platform, and a pipe fitter had fallen 80 feet into the sea, where he drowned. After consideration of various possibilities, it was decided to prosecute the installation's owner. The concession, however, was jointly owned by two multinationals, while the installation was registered in the name of only one of them. Whether it was uncertainty on this score or difficulty in building up the technical side of the case that delayed matters cannot be established. What can, however, is that as the time bar of six months for the commencement of summary proceedings approached,[75] the Procurator Fiscal had still not received final confirmation from the Department of Energy of the identity of the owner. As a result, he was having to contemplate recourse to an unusual Scottish procedure known as obtaining a 'warrant to cite', a device which permits the formal commencement of proceedings when the time limit has practically expired, even if the hearing of the substance of the case cannot be begun.[76] In the event, the company was prosecuted with a few days to go, and having pleaded guilty was fined £200.

The scaffolding company referred to in Case 7 had a contract to erect and maintain scaffolding on a major North Sea production platform. While one of its employees was standing on the scaffold, he stepped on to a plank which snapped beneath him. Remarkably, he survived the subsequent fall and a period of some twenty-seven minutes in the water, though not without injury. The major proposition underpinning the prosecution which followed was that the plank was rotten, a view supported by an expert report, which said that the 'signs of inevitable failure' were 'plainly to be seen'. How such an omission could arise within the complex network of contracts and subcontracts employed in offshore operations became apparent from a statement made by one of the company's employees, who explained that their job was to put up scaffolding when required to do so by other trades, but that once it was in place, they had nothing to do with it until they were asked to take it down. He stated that he had been aware that some of the boards used were inadequate and had complained to his employers about this, without any success. The company pleaded not guilty to the charge brought under Section 2 of the Health and Safety at Work Act, but was convicted and fined £400. For a time, it seemed that there might be further complications, inasmuch as an appeal to the High

Court of Justiciary was contemplated on the grounds that the company, having changed its name during tow-out might be a different legal person (different, presumably from the one which originally hired the injured person), but this course of action was abandoned.

With Case 8 we stumble across the name of one particular diving company for the third time. On this occasion, a winch designed for the launching and recovery of a Zodiac (an inflatable boat), had been involved in causing the non-fatal injuries sustained by two divers who had been dropped 50 feet into the sea. The winch in question was fitted with a disengagement mechanism disapproved by the Department of Energy (and by engineers in general) for purposes of lifting or loading personnel, and indeed, there was a manufacturer's warning to this effect on its side. As far as the accident itself was concerned, the proximate, technical cause appears to have been that the relevant part of the mechanism was gummed up with paint, a clutch lever therefore being enabled to move into the 'disengage' position. The need to dispel distressing on-board rumours about the incident, coupled with the fact that weather-dependent work was being delayed, led to the dismantlement of the winch for examination before the Department of Energy inspectors had completed investigation or given permission for this to be done. While the company was not prosecuted on this score, it was proceeded against under the Offshore Installations (Diving Operations) Regulations 1974 for failing to ensure that all plant and equipment necessary for the safe execution of the dive was provided, was of sound construction and suitable material, was in good order and was suitable for the purpose. The diving supervisor was also prosecuted under the same Regulations, and consideration was given to proceeding against the operating company, a multinational, as well. In the latter context, however, familiar problems arose out of the fact that the dive had been carried out from an accommodation vessel alongside the platform, doubts thereby being raised with regard to jurisdiction and attribution of ownership, not to mention the operator's status as employer. No further action was taken on this front, and instead the diving company and its supervisor were taken to court and fined £300 and £50 respectively.

With this case, the run of success which, with one exception, the authorities had enjoyed came more or less to an end, the tables now being turned, inasmuch as only one of the thirteen parties involved in the remaining five cases to reach court was convicted. In Case 9, the Department's doubts as to whether an offshore crane constituted plant or premises under Section 4(2) of the Health and Safety at Work Act seem to have contributed to an insoluble ambiguity in the wording of the complaint against the operating company, and this, combined with inadequate specification of wherein lay the lack of safety, was sufficient to secure dismissal of the charge on the grounds of irrelevance.[77] A member of the crane's commissioning team had been coming down off the crane, had missed his footing on the bottom step of a ladder and had fallen into the sea through a space previously closed off by a scaffolding platform which had been removed to facilitate heavy loading. But attempts to secure a conviction against the scaffolding company for failure to protect from risk persons *not* in their employment suffered a similar fate, and all that was salvaged was a conviction against the injured person's immediate employers for failure to provide adequate supervision.

Case 10 arose out of yet another accident to a scaffolder. This time, scaffolding had just been dismantled when a member of the work party fell 80 feet into the sea through a hole in a walkway where part of the grating had been removed. While it had been removed by the team itself, in order to pass tubes and fittings through the hole, no safety net had been provided. To his credit, the installation's manager, who had apparently been transferred to the platform following a fatality on another installation, had been dogmatic about the need for such netting, but his views do not seem to have prevailed. The injured person in this instance might presumably have wished it otherwise, being on his first offshore job and having come out to replace a scaffolder killed in a similar accident some time earlier. As the diet set for the case approached, however, the authorities became less and less confident about the prosecution's viability, not least because it might be argued that erecting such a net was just as dangerous. While the HSE was keen to proceed, the Department of Energy was not, and the case against both operators and scaffolders was deserted.

An all too familiar tale of haste taking precedence over safety lay behind the events leading up to another prosecution (Case 11). An attempt was being made to separate two joints of pipe while 'breaking out' (unscrewing) 20-inch casing. In order to save time, a shortcut involving the removal of chain wraps from around a cathead was taken, the wraps being put around the pipe itself and the chain's other end attached to a non-load-bearing connection on the cathead. When an effort was made to jerk the top joint of pipe by operating the cathead – which would normally apply frictional pull to a set of back-up tongs attached to the pipe – the connection broke and injured two workers on the drill floor, one of them quite severely. According to one report on the incident, it would only have taken an extra thirty minutes to utilize the proper equipment, while the training provided by the suppliers of the casing crew was described as 'similar to [that provided by] other drilling-related companies which prefer to get on with the job as quickly as possible'. Charged with failure to provide adequate instruction, information, training and supervision under Section 2 of the Health and Safety at Work Act, both the casing crew suppliers and the drilling contractors pleaded not guilty, while the case against one of the injured parties was deserted on the grounds of fairness, coupled with the more instrumental consideration that he would have to be prosecuted first and then used as a witness against the companies. In the end, both companies were found not guilty. The injured person had been experienced and had known the risk.

The penultimate case in what was now becoming a growing catalogue of failure was a fairly unmitigated disaster, to be outstripped only by the one which followed it. Following an injury sustained by a worker when a stabbing board, allegedly inadequately suspended, gave way and fell 20 feet, it was decided to throw the book at pretty well everyone concerned. The proprietors of the platform in question, the casing company which employed the injured person, the drilling contractors and the drilling supervisor were all charged under appropriate sections of the Health and Safety at Work Act. Initially, it seems, all of the accused were intent on pleading not guilty, but as time went on, other and potentially more effective strategies were to become available. Nearly five months had elapsed between the date on which the alleged offences were committed and that on which they were reported to the Procurator Fiscal by the Department of Energy. Left with only some five weeks to go before the procedings would become time-barred under Section 331 of the Criminal Procedure (Scotland) Act 1975,[78] with one day to spare, the Fiscal applied to the court for a warrant to cite. This was granted, and a diet just over three weeks ahead was assigned.

In the meantime, however, defence lawyers had been having second thoughts about the need to mount an elaborate case in support of a not guilty plea. Cognizant of the then pending decision on the relevance of the proceedings in another case (Case 9), those representing one of the accused companies intimated that they would now be entering a challenge on these grounds and asked that, when the case was called at the diet assigned, it should be continued without plea. This request was acceded to, and the case was put back to another date just under three weeks ahead. When the day arrived, challenges to relevancy were entered on behalf of all the accused, lack of adequate specification in the charges being claimed as the grounds. But a further escape route had opened up by this time. Following the granting of a warrant to cite, any delay in serving the complaint on the accused has got to be explained, and such delay there had been. The relevant complaints had been passed to the police for service after the warrant had been granted, but a fortnight had passed before this was done. This had left the accused with only a week's notice before the diet assigned at the first hearing, and, accordingly, when the case was resumed after its continuation without plea, further challenges were entered. This time they were against the competence of the proceedings on the grounds that the complaints had not been timeously served. The only explanation which the authorities could offer for the fortnight's delay was that the police had been hard-pressed at the time. After an adjournment to consider these submissions, the Sheriff ruled that the Crown had failed to show that there had not been undue delay in serving the complaint, that it had not been served timeously and that the charges were therefore time-barred. The complaint was duly dismissed as incompetent.

Finally, we come to a case which warrants somewhat more protracted discussion because it both represents the nadir of UK offshore safety law enforcement and encapsulates, in acute form, some of the other issues raised in this and previous chapters. The background to this case starts with a contract between Infabco Diving Services Limited and BNOC pertaining to operations connected with a riser on a single anchor leg mooring (SALM) in the Thistle field.[79] The work to be undertaken by the contractor (Infabco) included subsea diving, and the contract prohibited the employment of any subcontractor without the prior consent of BNOC. It also stipulated that the 'contractor shall observe and abide by and shall require its subcontractors to observe and abide by all laws, regulations and bye-laws as may apply in relation to the work under the Agreement'.

On 7 August 1979, work being carried out under the terms of the above contract necessitated a dive to a depth of some 490 feet, and this operation was commenced with the use of a diving bell manned by Richard Walker and Victor Guiel, two divers already in saturation. The bell was lowered over the side of the vessel rather than through its moonpool, and some time after it reached its designated position, the main lifting wire became disconnected. Although communications were lost around this time, the diving supervisor in the control room on the surface was able to ascertain that the one diver who had been outside had successfully re-entered the bell and had secured a seal on the bottom door. He also knew that the bell was now suspended on its umbilical, a collection of tubes and wires gathered round a central cable, and that it had slipped the remaining 30 feet or so to the sea bottom. When attempts were made to raise it by the umbilical, the components of the latter splayed apart and became entangled in the umbilical winch, with the result that the bell could be neither raised nor lowered – if, in fact, it had left the sea bed at all. Further efforts to take the strain off the umbilical by using a crane, so that repairs could be effected, resulted only in additional and severe damage. A number of the elements broke, and any thoughts of carrying out the rescue by using the crane to haul up the umbilical were abandoned, not least because it seemed that further pulling would have parted the latter altogether. In the end, and with the help of another diving vessel, the bell was located and brought to the surface some eighteen hours after the emergency had begun. Both of the occupants were dead.

En passant, it is worth pausing here to note that, despite all the drama of a deep-sea rescue bid, this accident bears all the hallmarks of the pattern described in chapter 3 of this book. Indeed, the entire incident could be described in terms of the technologically mundane. Quite apart from the fact that the emergency seems to have been precipitated by nothing more technologically sophisticated than a pin breaking or working itself free from the shackles attaching the bell to its main lifting wire (an eventuality for which some pretty commonsense explanations were offered at the subsequent trial), the other circumstances surrounding the tragedy scarcely smack of the inevitable contingencies to be expected at the very frontiers of knowledge. Umbilicals, for example, are regarded as an acceptable secondary means of lifting, provided that they are of a quality which precludes differential stretching of the component tubes and wires. This can be accomplished by sheathing the whole affair in a stocking of steel wire or nylon mesh. In the case in question, however, the trail-blazing technology of the North Sea had involved not sheathing but taping the umbilical's components together at 15-inch intervals with adhesive tape, ties which certainly did not prevent them from coming apart when the umbilical winch was called into action. Nor was the latter exactly the last word in modern engineering and design, as an expert witness told the court. Including car-type rubber tyres as essential components, one of which lacked sufficient internal support to withstand the weight put on it, the construction 'was not suitable for its reserved purpose as a means of recovery of the bell'. Even if some of the personnel involved thought that it had this reserve function, the Certifying Authority which had examined the diving system certainly did not. As a representative of Det Norske Veritas (DNV) testified, it had not been certified for this purpose.

All of this did not mean, however, that there was no alternative means of recovery, in theory at least. Indeed, in this context DNV had approved a system whereby ballast weights slung on the outside of the bell could be released from the inside, thereby creating the positive buoyancy which would bring it to the surface, albeit fairly unceremoniously. But the technological sophistication that had been brought to bear on the risk of such weights being released accidentally, a risk highlighted by the fatally rapid ascent involved in one of the accidents described in chapter 3,[80] testifies to the industry's ingenuity – they were lashed on with nylon rope. As the DNV witness told the court, the certificate in question specified that it would become invalid if 'essential alterations' were made to the system without approval, and, in his view, lashing the ballast weights to the bell constituted just such an alteration. Even then, however, all need not have been lost had it been possible for the two divers to cut the rope, return to the bell and, having sealed the door, release the weights. But the final and pathetic irony was that the bell had been designed for lowering down guide wires to rest some distance *above* a bottom or 'clump' weight, the fact that its door was more or less flush with the bottom of the bell frame not therefore impeding exit in normal circumstances. For undetermined reasons, the guide-wire system was not being used on this occasion, and the bell was simply suspended on its lifting wire and umbilical. Thus, when it went to the bottom, the flush door could not be opened sufficiently to permit both weights to be cut loose. The rope securing one ballast weight had been severed, and the other bore marks of possible attempts at cutting. While the truth about this aspect of the affair will probably never be known, the evidence would be consistent with a situation in which it had been possible to get an arm out far enough to cut one securing cord but not the other.

It was to this catalogue of cumulative catastrophe that an attempt was made to apply 'the full force of the law'[81] in December 1980, Infabco Diving Services Limited being proceeded against on indictment before a jury, and being charged with a series of offences under the Submarine Pipelines (Diving Operations) Regulations 1976. While the accident itself may have been disastrous enough, however, the ensuing trial turned the entire affair into a complete fiasco. Little more than a fortnight before the proceedings commenced, the Procurator Fiscal was appraised of the fact that Infabco intended to deny being the employers of the two men who had died, an ill omen for the trial inasmuch as each of the charges was prefaced by the words 'being the employer of'.[82] Presumably in order to retreive evidence which might be germane to this issue, he asked to see the contract between that company and BNOC, and, after some initial difficulties which almost resulted in the improbable spectacle of an application to the court for a warrant to search for and take possession of the contract in question, this document was finally obtained some days before the trial opened.

When it did, the issue soon became clear. Although there was some doubt as to whether Infabco had notified BNOC or not, it was their categorical claim that the diving side of the job had been subcontracted to another company known as Offshore Co-ordinators Limited, which therefore employed the divers. While at least some witnesses were under the impression that this was a Channel Islands company, in this they were technically mistaken, since subsequent investigations have shown that Offshore Co-ordinators (CI) was not registered in Jersey until 5 December 1979, some four months after the accident took place. A more plausible explanation is the one confirmed in evidence by Brian Masterton, one of the directors of Infabco Diving Services, namely, that this was a company incorporated under the laws of the Isle of Man and having an administrative office in St Helier, Jersey. Whatever the legal provenance of this company, however, the defence was astutely adamant in insisting that Infabco was not the employer and, no less important, in heading off the introduction of any documentary evidence to show that it had been acting in this capacity.

Particularly frustrating from the prosecution's point of view was the fact that on almost every occasion when witnesses were led towards stating the content of forms which might have purported to name Infabco as the employers, this was objected to on the grounds that they could not 'prove' the documents in question; that is, although they might have received a form, even from Infabco, they could not attest to the circumstances of its completion, and their evidence therefore amounted to nothing more than hearsay. Since the divers were dead and no one from Infabco could be produced to provide the requisite testimony, the Crown was in considerable difficulty. Thus, to take what was to become one of the key turning points in the whole trial, when a Department of Energy inspector was led towards revealing the content of Form OIR 9, an official casualty report which, among other things, requires the name and address of the employer to be filled in, the objection was that since he had not produced the document, he could not speak to its content. Even less could he be allowed to continue when he said that an Infabco official had handed him copies of the applications which the two dead men had made for employment. Although the Fiscal had to concede that in court documents cannot be allowed to speak for themselves, he nonetheless managed to get on to the record the question he wanted to put:

> What I was proposing to put to Mr Giles was: in response to a request for information furnished by Guiel for employment as a diver, did you receive from Infabco Diving Services Ltd, Production No. 17, which is headed 'Infabco Diving Services Ltd Application for Employment', and the name is Guiel, Victor F.?

For four days the increasingly unequal struggle continued, with the prosecution at times becoming almost desperate in its efforts to get crucial evidence before the jury. But to little avail. Even when successful, moreover, the evidence was easliy explained away. Although several pages in the logbook of one of the deceased were clearly stamped 'Infabco Diving Services', these stamps related to dates some time before the accident, and anyway the whole document emanated from yet another company – Infabco Diving Services Limited of Dublin, Eire. The British company had not employed divers since around February 1979, and there was no truth in the suggestion that the Irish organization, or Offshore Co-ordinators, was an arrangement for convenience only. Similarly, while letters signed on behalf of Infabco Diving Services and confirming the appointment of diving supervisors on the vessel in question were read to the court, this cut little legal ice. They were exactly what they purported to be – letters providing legally required, written confirmation of appointment to diving supervisors who were nonetheless employed by Offshore Co-ordinators.[83] In the end, the Sheriff brought the whole sorry business to a not unmerciful end by directing the jury to return a verdict of not guilty 'because of a failure to lead sufficient evidence on a fundamental fact in the case, namely, the question of the accused being the employer of the two divers'. Thus, although the fatal flaw in having Regulations which could not readily be accommodated to the complex and opaque employment relationships prevailing in the world of offshore diving had been known for some time,[84] it had survived long enough to sink the law's first really major foray into self-assertion almost without trace. Both evident and justified, the Sheriff's irritation provides a suitable note on which to conclude this chapter:

I should add that I hope that the Department of Energy will take account of the evidence given in this case about the employment position of divers in the North Sea and will consider, as a matter of urgency, whether regulations require amendment to widen the definition of 'Employer of diver' or in some other way. It is clearly undesirable that persons responsible for providing diving equipment should be able to escape the consequences of any breach of the regulations applying to that equipment, if such breaches occur. Whether they occurred in the present case is not, of course, for me to say.

Notes and References

1. For a discussion of the origins of 'conventionalization' with regard to the Factory Acts, see my paper 'The Conventionalization of Early Factory Crime', *International Journal for the Sociology of Law*, vol. 7, 1979, pp. 37–60.
2. See chapter 3 above.
3. A. Gouldner, *The Coming Crisis of Western Sociology*, London, Heinemann, 1970, p. 297.
4. See, for example, the classic work in this field by Edwin H. Sutherland, *White Collar Crime*, New York, Dryden Press, 1949.
5. W. G. Carson, 'White-Collar Crime and the Institutionalization of Ambiguity: the Early British Factory Acts', in G. Geis and E. Stotland (eds.), *White-Collar Crime*, Beverly Hills, Sage Publications, 1980.
6. J. Scott, *Corporations, Classes and Capitalism*, London, Hutchinson, 1979, p. 153.
7. See Carson, 'White-Collar Crime and the Institutionalization of Ambiguity'.
8. That is not to say, however, that the forging of such links is in itself sufficient to elucidate the relationship between enforcement practices and the totality. To grasp the full nature of the relationship, it would be necessary to show how law enforcement is not only shaped by but also serves to reproduce crucial features of the social order in which it takes place. This inevitably would involve, among other things, discussion of the ideological role of safety legislation and its enforcement, topics which cannot be addressed here. But for a discussion of how such an analysis can be brought to bear on the early history of factory legislation, see the paper referred to in note 5 above.
9. See chapter 5 above.
10. ibid., p. 44.
11. *Offshore Safety* (the Burgoyne Report), Cmnd. 7841, London, HMSO. 1980, p. 35.
12. *Ibid.*, p. 228.

13. *Parliamentary Debates* (Commons: Standing Committee G), vol. 4, 1970–1, 17 June 1971, col. 13.
14. For a discussion of these gaps, see the Burgoyne Report, p. 29.
15. Prohibition notices, applicable under the Health and Safety at Work Act, do not appear to be used by the Department of Energy, but the inspectors do have powers to order the cessation of activities where they apprehend the possibility of an accident taking place. See below, p. 247.
16. *Parliamentary Debates* (Commons), vol. 816, 1970–1, 28 April 1971, col. 664.
17. *Oilman*, 4 March 1978.
18. SI, 1973, 1842, Regulation 11.
19. ibid., Regulation 3(1)(g).
20. The investigation of fatalities and serious accidents does not, of course, involve the giving of such notice, although the inspectors are still dependent for transportation.
21. SI, 1976, 1019, Regulation 22(1).
22. This does not include accident investigations.
23. See chapter 5 above.
24. SI 1974, 289, Regulation 3; see also, the Burgoyne Report, p. 29.
25. The Department's letter to Certifying Authorities conceded that this might sometimes be needed and went on, 'as this Guidance Note is also being forwarded to licensees so that they are also aware of the guidance you have received, this is not being emphasized or even admitted.'
26. *Parliamentary Debates* (Commons: Standing Committee G), vol. 4, 1970–1, 17 June 1971, col. 12.
27. See chapter 5 above.
28. SI, 1973, 1842, Regulation 2(1)(g).
29. Burgoyne Report, p. 60.
30. ibid., p. 150.
31. Mineral Workings (Offshore Installations) Act 1971, c. 61 Section 7(3)(a). Imprisonment for up to two years, or an unspecified fine, or both is provided by Section 7(3)(b), which allows for proceedings to be taken on indictment. But this course is rarely followed in this as in other areas of safety regulation. Similar penalties are provided by the Health and Safety at Work Act 1974.
32. In 1977, for example, prosecutions for or by enforcement authorities coming under the Health and Safety Commission totalled 1720. *Health and Safety Statistics 1977*, London, HMSO, 1980, p. 7.
33. W. G. Carson, 'Some Sociological Aspects of Strict Liability and the Enforcement of Factory Legislation', *Modern Law Review*, 1970, vol. 33, no. 4, pp. 396–412.
34. Burgoyne Report, p. 60.
35. SI, 1977, 1232.
36. ibid., Art. 5(2)(ii); Mineral Workings (Offshore Installations) Act 1971, Section 12(2).
37. M. Scott, *The Racing Game*, Chicago, Aldine Press, 1968, ch. 2.

38.   See below, pp. 261–4, but cf. note 60.
39.   1964 UKTS 39; Cmnd. 2422. For a useful commentary on the Convention, one upon which the following discussion heavily relies, see the version annotated by J. P. Grant, in T. Daintith and G. Willoughby (eds.), *A Manual of United Kingdom Oil and Gas Law*, London, Oyez Publishing, 1977, pp. 167 ff.
40.   Section 9(5) of the Mineral Workings (Offshore Installations) Act 1971 similarly refers only to the 'powers, protection and privileges' of constables on installations.
41.   Burgoyne Report, p. 25. But for possible developments on this front, see *Draft Convention on the Law of the Sea* (informal text), September 1980, A/CONF./62/WP.10/Rev. 3, Art. 60 (5).
42.   Grant, in Daintith and Willoughby (eds.), *A Manual of United Kingdom Oil and Gas Law*, p. 177.
43.   ibid., p. 178.
44.   SI, 1977, 1232, Art. 5(1).
45.   See Burgoyne Report, p. 25.
46.   Section 20(2)(b).
47.   Burgoyne Report, pp. 124–5.
48.   See the report on the Conference's progress in the *Sunday Times*, 26 April 1981. John Grant, to whom I am indebted for his assistance on this point, confirms the tendency towards distraction from the problems with which we are here concerned.
49.   See chapter 6 above.
50.   Burgoyne Report, p. 125.
51.   Coroners Act 1887, c. 71, Section 3(1).
52.   58 & 59 Vict., c. 36; 6 Edw. 7, c. 35.
53.   See chapter 5 above.
54.   SI, 1968, 892.
55.   In this section, all references to cases in which no Fatal Accident Inquiry was held are correct as of July 1978, when the information was collected. Under retrospective provisions included in the Fatal Accidents and Sudden Deaths Inquiry (Scotland) Act 1976 (1976, c. 14), some of them may have subsequently been the subject of such an inquiry. See below, p. 262.
56.   Section 1(3).
57.   SI, 1972, 1542, Regulation 1(3).
58.   For a description of what is involved, see A. V. Sheehan, *Criminal Procedure in Scotland and France*, Edinburgh, HMSO, 1975, p. 224.
59.   Interestingly, according to one journalist who has worked on this aspect of the North Sea saga, a company official has subsequently admitted that the real worry in this case was that a Fatal Accident Inquiry might entail the release of operational information that the company preferred to keep confidential. *Glasgow Herald*, 24 June 1981.
60.   At least one eminent Scots lawyer still finds this loss of jurisdictional nerve suprising, since the terms utilized in Section 3(2) of the Continen-

tal Shelf Act 1964 are, in his view, quite wide enough to have embraced the holding of inquiries. (Personal communication from Professor R. Black, Department of Scots Law, University of Edinburgh.) Professor Black points out that the side-note to the section indeed referred only to the criminal and civil law. But side-notes have no legislative force and cannot be used as aids to construction. *Chandler* v. *DPP* [1964], A. C. 763 at 789.

61. SI, 1975, 1708. The main difficulty was that jurisdiction was still tied to 'acts or omissions', terms which might not, of course, cover all deaths occurring on the Continental Shelf.
62. 1976, c. 14.
63. Section 1(2).
64. Interview with Crown Office official, July 1978.
65. Under Section 1(1)(b), however, the Lord Advocate could still order an inquiry if the death 'occurred in circumstances such as to give rise to serious public concern.'
66. SI, 1968, 892, Art. 2.
67. Though Crown Office may be consulted and the instruction of Crown Counsel sought.
68. For a full list of the Procurator Fiscal's diverse functions, see Sheehan, *Criminal Procedure in Scotland and France*, pp. 110ff.
69. This misunderstanding frequently arises in connection with Scots law because of a stipulation that 'no person can be convicted of a crime or statutory offence, except when the legislature otherwise directs, unless there is evidence of at least two witnesses implicating the person accused. . . .' (*Morton* v. *HM Advocate* [1938], J.C. 50). As Sheehan points out, however, this does not mean that two witnesses are required to prove every fact in the case. 'It is sufficient if one witness is corroborated by facts and circumstances spoken to by others' (*Criminal Procedure in Scotland and France*, p. 121). Thus the problem is not quite as insurmountable as the inspector suggested.
70. See below, p. 268.
71. In Scotland, cases can be deserted (dropped) either *simpliciter* or *pro loco et tempore*. In the former instance, the effect is the same as a verdict of not guilty, no further proceedings being possible; in the latter, it is possible to start fresh proceedings at a later date, unless the case is time-barred as a statutory offence. See Sheehan, *Criminal Procedure in Scotland and France*, p. 159. On the question of time bars, see below, p. 274.
72. See below, pp. 275 ff.
73. But for other legal gaps pertaining to the construction stage, see above p. 235.
74. Fatal Accidents and Sudden Deaths Inquiry (Scotland) Act 1976, Section 5(1). The preservation of this power in the 1976 Act may well have been designed to cover just this kind of situation, where previously unknown facts emerge at an inquiry and point to a breach of the law.

75. Unless specifically extended by the relevant statute, this time bar applies to all summary proceedings arising out of statutory offences. Criminal Procedure (Scotland) Act 1975, Section 331.
76. See Sheehan, *Criminal Procedure in Scotland and France*, p. 140.
77. Objections to the relevancy of proceedings on grounds such as these are quite common. In this case, the main problem arose from the fact that under Section 4(2) of the Act duties towards non-employees depend upon the person in question having, at least to some extent, control over premises, access thereto or plant within such premises. Since the complaint did not aver that the operators had some control over the crane, but simply described it as 'on' the premises, it failed to make adequate distinction between unsafe premises (for which the accused might have been responsible) and unsafe plant (for which they might not).
78. See note 75 above.
79. For a brief description of this and other technical terms involved in the following discussion, see p. 306 below.
80. See pp. 67 ff. above.
81. See p. 152 above.
82. Regulation 5 lays a variety of duties quite specifically upon 'employers of divers'.
83. Regulation 5 requires 'employers of divers' to appoint in writing the persons who will carry out the duties imposed on diving supervisors by the Regulations.
84. See above, p. 270.

8

Conclusion

Truth and oil always come to the surface.
Spanish proverb

The first thing for a conclusion to do is to bring the story, if not to an end, at least up to date. A great deal has happened during the period in which the main body of this book has been in the writing; indeed, throughout that fairly protracted exercise I have been somewhat haunted by the possibility of being overtaken by events which might undermine my thesis at one or more vital points. Two official reports dealing with major oil disasters outside the British sector of the North Sea have been published; decisions on the implementation of the Burgoyne Committee's recommendations have been taken; and new diving regulations which close the loophole concerning the 'employers of divers' have been issued. With regard to the accident in connection with which the latter question became central to the criminal trial discussed at some length in chapter 7, a Fatal Accident Inquiry is in progress at the time of writing. And, of course, North Sea operations have gone on apace, with Britain's oil continuing to add yet more twists to her position within the world economy.

Tracing these developments through in the order suggested by their relevance to the content of previous chapters, the first point to be made is that, as suggested earlier, the increased number of reported fatalities during 1979 does seem to have been a statistical aberration.[1] In 1980, four deaths on installations or vessels coming under the Offshore Installations (Inspectors and Casualties) Regulations 1973 were reported to the Department of Energy, as against ten in the previous year.[2] Although it may be tempting fate even at this late stage, it does seem likely that the risk of death as a result of employment in these aspects of offshore operations is on the decline (but see p. 5). Unfortunately, the Department's adoption of new methods for calculating numbers employed precludes any attempt to support this view with incidence rates, while, as ever, these figures do not take account of any fatalities which may have taken place on pipelaying barges.[3]

On the assumption that the fatality figures for 1980 do reflect an underlying downward trend, where does this leave my argument about the dangers of offshore employment? Not seriously imperilled is the immediate answer, inasmuch as the bulk of this book has been about the other price of Britain's oil during the earlier and crucial period of development up to the late 1970s. Moreover, while incidence rates are again unavailable, the statistics for serious accidents and dangerous occurrences during 1980 suggest that any optimism should be tempered with a degree of scepticism. The number of serious accidents reported remained relatively stable at forty-five in that year, while dangerous occurrences increased from ninety-three to 118.[4] Even if the latter figure may be due partly to improved reporting standards, as the Department suspects, there is still justifiable cause for caution and concern.[5]

Nor is this the only reason for placing a guard upon our optimism. In recent months, serious doubts have been raised about an aspect of offshore safety with which this book has been able to deal only cursorily – the provision of stand-by vessels. At one level, it has been suggested that safety standards on these vessels themselves may be unacceptably low, while there have been allegations that the Department of Trade's code of practice is regularly contravened in several respects.[6] Parenthetically, it is worth adding here that the absence of any statutory regulations in this context serves to underline the fact that the legislative response to offshore operations has not even yet caught up with some fairly vital issues. Even more alarming, however, is the alleged deficiency of such vessels in another respect, namely, their capacity to make a significant rescue contribution in the event of a really major accident. Such an eventuality has not yet arisen in the British sector, and the Department of Energy accordingly has to admit to ignorance on this score. But the fact that the vessel on stand-by duty when the Alexander Kielland turned over retrieved two bodies and no survivors suggests a rather chilling hypothesis in this respect.[7] Worries about this question became more than hypothetical when, in May 1981, an exercise specially laid on for the benefit of two Department of Energy inspectors resulted in a fatality. The mate of the stand-by vessel fell overboard, and neither it nor another stand-by ship which went to the scene could effect a successful rescue.[8]

Nothing that has transpired during the writing of this book leads me to revise my views on the relatively conventional and preventable nature of most offshore accidents. While advanced technology and adverse operating conditions cannot be denied, I remain convinced that the source of the greatest danger in the North Sea is not to be traced to the unknown or the untamed. Here, of course, the major potential exception has been the calamity which overwhelmed the Alexander Kielland in March 1980, when 123 men died in a disaster that captured horror headlines throughout the world. Surely, I have been thinking, when one of the five giant floating legs supporting a structure approximately the size of the Albert Hall falls off, we must finally be confronting the unknowable contingencies of technology's frontiers or forces of a ferocity beyond the capacity of human ingenuity and legal codes. For close on a year, I have been writing with continual glances over my shoulder at the progress of the Norwegian authorities' investigation, which might show this accident to have been an exception challenging, if not completely undermining, my argument in relation to the British sector.

In April 1981, the 360-page report of the official inquiry into this disaster was published, and although the privately produced English translation is available only at a cost of more than £1000, an English summary of the findings, coupled with the kind assistance of some Norwegian friends, permits the main conclusions to be outlined.[9] The immediate cause of the accident was plain. Into an opening cut in one of the six bracings holding the relevant column (leg) to the platform a support for a hydrophone positioning control had been welded. This welding was 'not the best', and the fact that the hydrophone support would now comprise part of the load-carrying structure, as opposed to 'outfit' (equipment), had not been considered at the planning stage. Fatigue fractures had developed in both the welds connecting the hydrophone support to the bracing and in the support itself. Some of these fractures must have developed before the platform was placed in position because, among other things, paint residues from the Dunkerque construction yard where the Alexander Kielland had been built were subsequently found on fracture surfaces. Neither control over design planning nor surveys carried out during building and operation had been sufficient to reveal these defects.

On the day in question, the weather had been bad, though not, it may be added, particularly ferocious by North Sea standards. Wind velocity had been 35–45 miles per hour, and wave height 20–25 feet. By this stage the fatigue fractures mentioned earlier had produced a redistribution of stress, initiating further fatigue cracks in the bracing to the point where two-thirds of its circumference was affected. When the weakened bracing broke, its five companions were subjected to overloading and gave way in rapid succession. The column became detached from the platform, and the Alexander Kielland heeled over to an angle of 30–35 °, largely because neither design nor regulations had allowed for the need to maintain stability in the event of one of the five columns breaking loose. Even then, disaster might have been averted if there had been compliance with instructions pertaining to watertight doors and ventilators on the deck. However, with some possible assistance from holes caused by damage already sustained, these openings allowed over 50 per cent of the deck volume to flood, and the whole structure turned over completely in the space of some twenty minutes. This drastically reduced the time available for any attempt at rescue from the installation itself, and apart from those who remained to go down with the rig, the rest had to face the sea. As the report concluded, 'The chance of surviving more than half an hour in the cold water was, as it turned out to be, minimal.' Out of a total complement of 212, only eighty-nine survived.

In all of this, I would maintain, there is very little to contradict my central thesis about offshore accidents. To be sure, the scale of the disaster outstrips by a long way anything which has thus far taken place in the British sector, but the chain of precipitating factors is depressingly familiar. While underwater inspection to detect weld failures or cracks may indeed by extremely difficult, technological ingenuity is scarcely taxed if, as would seem to have happened in the present instance, they are there to be found at the pre-installation stage.[10] Similarly, failure to allow for the loss of a crucial buoyancy column must surely strike even the layman as analogous to designing an aircraft without considering the implications of the failure of one of the engines. One expert who commented on the safety of Pentagone rigs like the Alexander Kielland in the wake of the disaster used the same analogy to even more powerful effect when he observed that although the design was basically good, if it was not made structurally sound, the consequence was like 'building an airplane without making sure the wings stay on'.[11] Failure to close watertight doors speaks for itself.[12]

Clearly, then, we must be careful not to be misled by the sheer scale of disaster or by the apparent complexities of its provenance. The factors leading to the Alexander Kielland disaster were knowable; their remedy lay within the range of existent technology; and, no less important, they were almost certainly avoidable if a properly constructed, efficiently administered and faithfully adhered to system of legal regulations had been applicable. Moreover, I remain un- repentantly of the view that the British authorities' complacent belief in the superiority of their own regulatory approach will retain even the semblance of credibility only as long as a similar tragedy does not overtake an installation operating in the British sector of the North Sea.[13] Nor does the evidence adduced by another official investigation, this time into a tanker fire and explosion which took place in Bantry Bay, Ireland, with the loss of fifty lives, breed any confidence that the industry's good-will and sense of responsibility can be relied upon to prevent a similarly catastrophic concatenation of mundane and know- able forces in the future.[14] Quite apart from forthright comments on the 'steps taken to suppress the truth' by some Gulf personnel[15] and on the all too familiar tale of 'a highly anomalous legal situation' which meant that no relevant regulations had been made before the accident took place,[16] the Irish report into this disaster was commendably blunt:

(a) The seriously weakened hull of the vessel was the result of deli- berate decisions taken at different times by the management of Total. [These decisions were taken because it was then considered that the ship would be sold in the near future, and in the interests of economy. Inadequate consideration was given to the effect they would have on the safety of the vessel.] Neither the Master nor the Chief Officer could have been aware on the night of the disaster how seriously weakened the vessel was. . . . Had Total supplied the ship with a loadicator – as it should have – the ballasting error would not have been made. The major share of the responsibility for the loss of the ship must lie on the management of Total.

(b) Had the dispatcher in the Control Room observed the initiation of the disaster, it is probable that the lives of both the jetty crew and those on board the ship would have been saved.

(c) Had Gulf maintained the stand-by tug close to, and in sight of, the jetty, it is probable that, notwithstanding the absence from the Control Room of the dispatcher . . . the lives of the jetty crew and those on board the vessel would have been saved.

(d) Had Gulf supplied suitable escape craft at the jetty, it is probable that . . . the lives of the jetty crew and those on board the vessel would have been saved.

(e) Had access to the sea from Dolphin I been maintained and had the jetty crew been properly trained in emergency procedures . . . then it is possible – but no certain conclusions on this point can be arrived at – that the jetty crew on the centre platform would have been saved.

(f) Had the decision to discontinue the automatically pressurized fire-main not been taken, it is possible – but again no certain conclusions on this point can be arrived at – that the jetty crew might have been able to contain the fire. . . .

(g) Had the alert been raised at the beginning of the disaster or had the stand-by tug been closer to the jetty, it is probable that the fire would have been contained. . . .

(h) Had the tug been moored in sight of the jetty and close to it, it would have been able to contain the fire and probably extinguish it before it spread on either side of the ship. . . .[17]

Back in the North Sea itself, another major development has been the reorganization of the offshore safety regime's administrative structure following the publication of the Burgoyne Report. Announced to the House of Commons in November 1980, these changes included the transfer of virtually complete responsibility for offshore safety to the Secretary of State for Energy, the one major exception being control over safety on ships, which was to remain with the Department of Trade.[18] In addition to retaining front-line responsibility for enforcement of the Health and Safety at Work Act 1974 as regards offshore installations, the PED was to take over the task of implementing that statute in connection with pipe-laying activities. Moreover, while the first of these arrangements would still entail an agency agreement with the HSE, that body's answerability in this context would now be to the Department of Energy rather than, as formerly, to the Secretary of State for Employment. Although the HSC would assume responsibility for advising on health and safety policy, including advice on blow-outs, associated fire prevention and technical aspects of installations' structural integrity, this apparent compromise was heavily qualified in two ways. The input of actual policy work to the Commission would come from the PED, and the Commission's advice would in turn be tendered to the Secretary of State for Energy. To strengthen the PED in both policy and enforcement matters associated with occupational health and safety, four experienced inspectors were to be transferred to that body from the HSE.[19]

The announcement of these proposed changes sparked off a stormy debate in the Commons. On one side, the Minister of State at the Department of Energy based his case on the by now familiar argument which claims offshore operations to be specialized, and organized in a way so unlike anything on land, that an administrative exception to the general rule was fully justified. Inevitably, the argument was embellished by allusion to the vast scale of offshore installations and to wind and wave forces which 'can be among the worst in the world'.[20] While the Burgoyne Report's Note of Dissent calling for the transfer of responsibility to the HSE had been fully considered, it was only a minority view and had not been found persuasive.[21] In terms of the argument advanced in chapter 5 of this book, it is also interesting to note that one of the main reasons for rejecting the dissenters' recommendation was that such a step would divorce the PED's inspectors from 'immediate and continuing access direct to other petroleum specialists in my Department', specialists who 'must be retained in my Department because of the nature of their main responsibilities for field developments and operations'.[22]

Opposition to the proposals followed thick and fast, and not just from members of the Opposition itself. While not everyone was persuaded that the HSE's attitude to enforcement was entirely above reproach,[23] the generally accepted view was that transfer of responsibility to that organization was necessary in order to separate the administration of safety regulations from the departmental role of sponsorship. Several speakers rehearsed the history of how such separation had been found essential and acceptable for other highly specialized industries in the past, and some could not hide their suspicion that the intransigence being encountered on this occasion was really 'to do with the preservation of the empire of the Department of Energy and [its] vested interests'.[24] What would happen, asked one Member, with deliberate exaggeration, 'if there were a major shutdown of oil supplies in the Middle East and a serious safety hazard was likely to develop in the Forties platform, and the United Kingdom was facing the problem of the supply of oil . . .?' Might the Department not 'accept the risk and keep Forties in production, regardless of safety?' Might not the Health and Safety Executive take the contrary view?[25] In more measured terms, the issue at stake was outlined succinctly by Mr. Robert Maclennan as being 'whether it is appropriate that the sponsoring Department of the industry, whose prime concern must necessarily be the maximization of production and the protection of that industry as it perceives its interests, should be responsible for safety'.[26] Echoing the views of almost all of the speakers, as well as one of this book's central arguments, Mr Ernie Ross summed the whole position up with the help of a story:

> It is interesting to note a statement made recently by the Secretary of State for Energy at a Texaco dinner. He went out of his way to congratulate the multinational oil corporation Texaco. He gave the game away to many of us when he said that 'The development of our

North Sea oil reserves has taken great skill and massive investment,
and has probably been achieved faster than in any other oil province
in history'. That is exactly what concerns Labour Members in this
debate. We are concerned about the speed with which the oil companies wish to extract this vital resource from the North Sea, and
with the direct involvement that the Department of Energy has with
them in ensuring that the rate of extraction is maintained. We are
concerned that they should be responsible for safety in the North
Sea. That is why we would certainly support the recommendations in
the minority report calling for a separate organization. We feel that
that organization should be the Health and Safety Executive.[27]

Despite the overwhelmingly unfavourable reaction from nearly
everyone who spoke in this debate, an amendment designed to give
formal expression to the view that responsibility for offshore safety
properly belongs to the HSE was lost by sixty votes to ninety-nine. As a
consequence, it can be said that the forces which were described in
chapter 6 as permeating the internal politics of the state have indeed
substantially restored the Department of Energy's monopoly on
offshore safety, and that this area of safety regulation will therefore
continue to be, at least for the immediate future, a rather glaring
anomaly within the broader field of British legislation on industrial
safety. What will happen in the longer term is, of course, impossible to
determine. On one reading of the situation, the HSC and the HSE are
now either satisfied with, or reconciled to, the new arrangements and
will be content to carry out their revised functions for the duration of
the period in which Britain obtains oil and gas from her own Continental Shelf. Indeed, some support for the view that internecine wrangling
within the bureaucracy has now come to an end can be derived from
the fact that I have been asked to stress that some of the views
attributed to officials of those organizations at earlier stages in this
book do not represent their current thinking. On the other hand, it
must be pointed out that the parliamentary debate on the Burgoyne
Report did demonstrate fairly firm commitment on the part of the
Labour Party to reversing the new arrangements if it should regain
power. In such an eventuality, moreover, the fact that HSE inspectors
will by then have been working within the Department of Energy for
several years would obviously have some effect in facilitating the
transfer of effective control to the Executive. The history of conflict
over responsibility for North Sea safety has not, I suspect, come to an
end with the implementation of the Burgoyne Report.

One factor which the protagonists in the struggle over the latter decision could scarcely have taken into account was what was going to happen to the pound within a very few months of their deliberations. By early June 1981 what some see as sterling's 'silly elevation to petro-currency status'[28] had allowed the currency to enjoy an eighteen-month period of relative insulation from the growing turbulence of the international money markets, largely stirred up by high US interest rates and the consequent strengthening of the dollar. During the same period, however, a glut in world oil supplies had been developing, partly because of reduced demand and partly because of Saudi Arabia's willingness to maintain high production levels in order to stem price increases. When Mexico cut the price of her crude early in June, several things followed in rapid succession. First, sterling's attractiveness as a petro-currency came under sharp re-evaluation, and the pound crashed back down through the $2 barrier to take up an uneasily stable position (perhaps temporarily) at around $1.94. Under increasing pressure because North Sea prices were tied to the higher and now unrealistic prices charged by Libya, Algeria and Nigeria, BNOC offered to reduce the price at which it sells its 51 per cent share of North Sea oil back to the companies by $2 per barrel. In turn, the companies sought a bigger reduction, of around $5 per barrel, started to move towards increasing the price of petrol at the pumps and threatened production cutbacks unless offered a deal which would stem the allegedly rising tide of losses being made on North Sea operations.[29]

What effect all of this may have on the British economy is impossible to predict, though it is certainly tempting to pursue the deluge of instant punditry on the subject in terms of the conjunction of forces discussed in chapter 4 under the rubric of combined and uneven development.[30] Nor is it possible at this stage to be clear about the possible implications of these recent changes for offshore safety. One possible scenario is of course, that further North Sea development might lose much of its urgency, while production cutbacks could be welcomed by the belatedly conservation-conscious lobby within the Department of Energy. If this were to happen, safety might move further up the scale of official priorities, and the ultimate removal of responsibility for this aspect of North Sea operations from the Department of Energy might be facilitated. On the other hand, the Treasury's interests in revenues – which are estimated to drop by some £180–200 million for every dollar's reduction in the per barrel price of North Sea oil[31] – could well generate pressure to keep prices and/or production prospects as high as possible. Nor should we forget that just as the external exigencies of the complex world oil economy triggered the events which have been described above, so they could just as easily reverse the situation once again. A change in Saudi Arabia's stance, or reprisals in the wake of Israeli attacks on Iraq's nuclear reactor could, for example, turn glut into the renewed spectre of shortage almost overnight. If this were to happen, North Sea exploration, development and exploitation would probably once again assume the character of a race against time, and the structural inadequacy of the revamped regulatory regime would rapidly become apparent once more. Whatever happens, however, the bulk of this book has been about the impact of speed on the regulation of offshore safety in the past, and while the course of history can indeed alter in the future, the record of what has gone before will retain its tragic obduracy as a matter of historical record.

On the statutory front, the most important development to have taken place during recent months has been the promulgation of new, unified diving regulations.[32] Drafted by the HSE and the Department of Energy's PED – the two bodies being, for this express purpose, kept 'in harness for the time being' under the revised administrative arrangements described above[33] – these regulations laid down crucial provisions for, among other things, ensuring the adequacy of divers' qualifications and closing loopholes pertaining to legal responsibility in the event of violation. Had these regulations been in force when Richard Walker and Victor Guiel were killed in August 1979, the trial which subsequently foundered on the latter issue would probably at least have proceeded to a more definitive outcome (even if not necessarily, I hasten to add, to a finding of guilt against the parties accused). That the relevant regulations were not in existence at the operative time serves to underline two issues discussed earlier in this book – the laggardly pace of the regulatory response to offshore dangers, and the way in which HSE involvement touched vitally raw nerves. On the first of these scores, it is relevant to point out that as early as November 1976, officials of the HSC explained in interview that diving was the one subject being looked at 'urgently', with the objective of producing a single diving code as soon as the Order in Council extending the Health and Safety at Work Act to the Continental Shelf went through.[34] The relevant Order came into operation in September 1977, and in April 1978, the Department of Energy stated that the HSE now had proposals for the regulations in question.[35] That the regulations in question did not finally emerge for almost another three years may not be unrelated to the second point. According to a reliable source, one of the main stumbling blocks impeding progress, as late as the summer of 1980, was that the HSE felt it was operating with 'a pistol to its head', inasmuch as any insistence on involving the trade unions in approval procedures might mean forfeiting the support of the relevant sector of the industry.

As for the Infabco saga itself, the accident in which Richard Walker and Victor Gueil were involved is currently the subject of a Fatal Accident Inquiry. While the final determination in this case is still some way off, however, reports of the proceedings are already revealing an extraordinary catalogue of alleged deficiency, chaos, misunderstanding and acrimony. A Department of Energy inspector, for example, told the inquiry that the Department had not been notified of the intention to carry out diving operations of the type envisaged, as it should have been; while, at least in his view, the reason why the two trapped men asked at one point to be placed back on the sea bed could have been that they were 'concerned with the strength of the umbilical or the durability of the umbilical to withstand the strain'.[36] If such indeed was their concern, the testimony of another inspector would seem to confirm that they were justified, since he did not believe that 'the use of the umbilical, particularly of that design, was adequate as a secondary means of recovery'.[37] According to another witness, the load line in the bell's umbilical cord had a breaking point of under 4 tons, less than half of its prescribed lifting weight of 10 tons, a fact which he and his colleagues had established by tests after the vessel finally docked at Lerwick.[38] As for Offshore Co-ordinators, the firm which had a contract to supply Infabco with divers, the Sheriff was reported as commenting on the fact that not only did this organization seem 'to consist of a lady who worked for a bank in Jersey', but also that they 'must have built up their expertise in the space of twenty-four hours', since it 'appeared that the contract had been signed the day after Offshore Co-ordinators were set up'.[39]

Accounts of the rescue operation itself scarcely do anything to dispel the general aura of chaos surrounding the tragedy. Worried about the strength of wire to be used in a proposed rescue line, one witness explained, he had gone to the stores for stronger wire but had found the door locked. After getting someone to open it, he took two lengths of wire to the deck and said he would be back shortly with two others. But despite his instructions to the contrary, the door had been locked again, and by the time he had eventually obtained the extra length of wire and had returned to the deck, the operation had gone ahead with a rescue line incorporating weaker wires.[40] More generally, witnesses told of confusion over who was in charge of the rescue attempt, of rows between rescue teams and even of extraordinary threats being made against one group of divers who, dissatisfied with how the operation was being handled, had wanted to take over. 'We were told we could be in very deep trouble,' reported one of them, 'We had to consider the possibility of a mutiny charge on the high seas.'[41] All in all, it is not perhaps inappropriate to conclude discussion of this particular accident by citing evidence from the only North Sea worker whose death has been covered in this book who left behind some record of his own feelings. Earlier in the day on which the fatal dive began, Richard Walker made the following entry in his diary about his personal troubles:

> On location at Thistle. The boat heaves a lot. There are conger eels all over. We will find out today. [No reference to what he would find out] Poor topside management. Guys here are nuts (ungood nuts) and dear God I want out. I have really got to scare up another job after this one. I'm no longer impressed. They have made no effort to get out gear from the X to dive first. . . . Not impressed. Too many Brits/political hassles. It just leaves my stomach twitching. . . . Oh, God, please help me to exercise my talent and will to pull out of it. I don't even know if I'm gonna get out of here alive. I never know.[42]

At the beginning of this book I endorsed an old-fashioned view which sees one of the essential elements of the sociological imagination as being the capacity to connect personal troubles with public issues. Moreover, I suggested, the forging of such links entails connecting personal troubles both to the major transformations that overtake particular societies and to the 'big ups and downs' which occur on a broader world scale. Indeed, it was argued, only when the dynamic interplay between global forces and the historically specific, unique features of the British nation state is teased out does it become possible to locate in the course of world history the personal troubles endured by a Richard Walker or by any of the other men whose deaths are now anonymously enshrined in the statistical records of North Sea casualties. Invoking the idea of combined and uneven development, I suggested that this approach offered the most useful analytical entrée to the empirical locus within which to seek the public issues connected with the personal troubles encountered by those who have died or been injured in the British sector of the North Sea.

It is with the charting of connections in this way that much of this book has been explicitly or implicitly concerned. Having mapped out how I see the complex intersection between developments within the world economy, its oil sector and the unique dimensions of Britain's predicament from the early 1960s onwards, I went on to argue that this concatenation of forces generated and sustained an overriding impetus towards the speedy development of the Continental Shelf's petroleum resources. In turn, this preoccupation with speed pushed the legislative response pertaining to the safety of offshore operations very much into second place and spawned a regulatory regime which became as special as it was inappropriate. Among other things, I argued, the oil sector of the economy was substantially able to 'privatize' one crucial part of the public administration. When this happens, however, competition between different social interests becomes displaced into the state apparatus itself, and in chapter 6 I examined some of the conflicts which ensued in the case of North Sea oil operations and their safety. On the one hand, the forces pushing for incorporation of offshore safety into the broader generic machinery of onshore safety administration enjoyed some temporary success in breaking the Department of Energy's effective monopoly in this sphere; on the other, the implications of such a move in terms of de-privatization, the enhancement of worker influence in a labour context where it is uniquely low, the possible compromising of Britain's international stance on her oil policy and, not least, the potential separation of safety considerations from the requirements of production all meant that this success was to be short-lived. Finally, I attempted to show that the same combination of factors which were so important in shaping legislative responses, regulatory relationships and the outcome of the state's internal wranglings over North Sea safety also found expression in the practice of law enforcement itself.

This is what I have tried to do, and I hope that, in some measure at least, my efforts have been successful. But now I find myself in something of a quandary. If personal troubles like those endured by offshore workers do have their genesis in structural transformations, in 'big ups and downs' and the rest of it, what is there to be done? Surely the forces which have so frequently figured in the pages of this book are of such massive dimensions and of such an inexorable character that there is little point in making piecemeal suggestions or in counselling changes which stop short of basic alterations to the entire social and economic system in which we live. Surely it is only thus that we could develop a society which would finally put the price of men's lives above the price of oil. Surely to call for anything less than basic transformation is simply to volunteer for the ranks of those who are pejoratively described these days as 'cautious rebels' or, apparently worse still, 'mere liberals.'[43]

Let me make no secret of it: a drastic transformation in the nature of the capitalist system might not, in my view, be any bad thing. Good or bad, however, such a transformation still seems a long way off, and short of the ultimate holocaust which would consign the North Sea to a totally oblivious history, the capitalist system seems set to outlast the resources of Britain's Continental Shelf by quite a long way. In the meantime, it seems to me that it would be both morally and intellectually irresponsible to dodge the issue of what steps might be taken to improve the offshore safety regime in the foreseeable future. To be sure, as I suggested above, this does not mean that I am so naive as to believe that law can offer a complete panacea to social ills such as the catalogue of death and injury which has been compiled in the course of offshore operations. Nor do I necessarily accept E. P. Thompson's famous assessment of the rule of law itself as 'an unqualified human good'.[44] With qualifications, however, it does strike me as the best thing we have to be going on with in the meantime. 'Anodyne' is a word too frequently used disparagingly; even palliatives cannot be responsibly ignored while final cures, if that is what they turn out to be, are awaited. In short, it is my belief that whatever space exists for immediate and effective action to improve the offshore safety regime has got to be occupied and resolutely defended even this late in the day.

From what has gone before, readers will probably not be surprised that my first suggestion is the transfer of front-line and policy responsibility from the Department of Energy to the HSC and the HSE. In making this proposal, I am conscious that I am advocating a greatly increased role for an organization of whose principal operational arm I have not been uncritical in the past.[45] However I repeat what was said in the introduction: the best should not be allowed to become the implacable enemy of the good. I am persuaded that changes of the kind proposed here would greatly improve the standard of offshore safety, even if leaving it far from perfect. Only by this means, I maintain, can even the necessary minimum of independence from the constraints of economic exigency be guaranteed. Transfer of requisite personnel, it seems to me, should not be allowed to present an insuperable problem, while such a step would mean that both specialist safety expertise and control over the mundane risks associated with North Sea operations were located within an organization less infatuated with the uniqueness of the offshore industry. As for access to experts in other aspects of North Sea operations, it does seem slightly specious to suggest that the mere fact of administrative separation would rule this out in the event of a major emergency, the kind of circumstance in which Government envisages it as being necessary.[46]

As suggested in chapter 6, changes of the kind proposed here have implications for the vexed question of trade union activity offshore. Whether or not my recommendation for a transfer of administrative responsibility is accepted, however, there is an urgent need for extension to the Continental Shelf of the Health and Safety at Work Act's provisions pertaining to safety representatives and safety committees.[47] This question has already dragged on for an unconscionable time, and the impression is growing, at least in my own mind, that Government is quite happy that problems arising out of the complexity of offshore employment practices should continue to impede progress on this front.[48] Nor do I think it would be sufficient to follow Burgoyne's line in this matter and extend the regulations in a modified form, which would exclude the necessity for union involvement. As suggested earlier in this book, it is by no means certain that the atmosphere prevailing on North Sea oil installations is such that the individual worker, even if elected by his fellows, would feel sufficiently secure to fulfil his alloted role without fear of consequences.[49] While health and safety affairs are often ideologically represented as a matter of men of good-will simply deciding what is best to be done, it should not be forgotten that here, as elsewhere, power intrudes. In any event, if the participation of the trade union movement is accepted as an integral part of onshore safety administration, I can see no *prima facie* reason for making an exception in the case of the North Sea. While the question of offshore unionization has not been one of this study's central concerns, a further implication of the above proposal is that, even purely on the grounds of safety regulation, Governments should in future take much more active steps to satisfy themselves as to the adequacy of the opportunities afforded for legitimate trade union recruitment. If the provision of access for this purpose is a condition of the licence, then there is a responsibility to ensure that this condition has been satisfactorily fulfilled.[50]

With reference to another matter of considerable practical import, the operational intersection between the legal systems of England and Scotland, the evidence adduced by this research suggests that confusion, misunderstanding and inadequate co-operation between the two jurisdictions have created serious problems in the past. Furthermore, and although it is not perhaps too much to hope that most of the major difficulties in this respect have now been surmounted, I am by no means certain that London-based decision makers are yet sufficiently conversant with the ways of the legal system under whose auspices so much of the law relating to offshore safety has to operate. For this they are not, in a sense, to blame. Scrutiny of the legislative debates discussed at an earlier stage in this book reveals more than one occasion on which legislators themselves gave a lead in conceding ignorance of Scots law and in treating it as the legal analogue of the stage Irishman. As we have seen, however, the consequences have sometimes unfortunately turned out to be more serious than mere offence to 'a curious set of laws', 'indignation being aroused in the corridors of Parliament House in Edinburgh and in the columns of the *Scots Law Times*', or 'a pathetic picture of the gentleman off Land's End, in some difficulty with his pint of beer, getting arraigned in Edinburgh'.[51] Offshore safety warrants less levity than that, and active steps should now be taken to ensure than any residual confusions and misunderstandings about the *modus operandi* of the Scottish legal system in this field are finally resolved.

The fact that so much of Britain's oil is recovered from an area falling within the jurisdiction of the Scottish courts also serves to underline a point of more general application, namely, what Edwin H. Sutherland once referred to as 'the bias involved in the administration of criminal justice under laws which apply exclusively to business and the professions'.[52] In Scotland, although the police have some small measure of discretion in the matter, the general rule is that they must report to the Procurator Fiscal all cases of 'ordinary' crime where there is sufficient evidence to justify taking proceedings against a particular accused.[53] However, agencies like the Department of Energy's PED or the HSE are not under the same obligation, and the question therefore arises of whether there is not an institutionalized discretionary bias at work in favour of offenders against regulatory codes such as those pertaining to safety in the North Sea. In the course of this research, it has not been possible to determine what proportion of observed contraventions is reported to the Procurator Fiscal for his decision on the question of prosecution, but the information presented in chapter 7 suggests that it is probably very low. If such is the case, it is not something about which independent research can do very much. Nor, indeed, may it be possible for the Scottish authorities to exercise much influence or make demands upon powerful and autonomous government Departments. What can be done, however, is to initiate a public debate on the issue of the equity with which the criminal justice system of this country is administered. If saying this consigns me to the company of those whose criminology was once castigated for being based on 'a mindless and atheoretical moral indignation',[54] I can only say that I hope there will always be room for some moral outrage in criminology and sociology alike. Moreover, I am consoled by Thomas Mathiesen's theoretically sophisticated and politically practical reminder that if an 'alternative' is to *compete* with an established system it must be presented in terms which are relevant to members of that system, in terms which cannot be disregarded as totally foreign, permanently outside and of no concern to the system itself.[55]

To raise questions such as these is to touch on the broader contradiction represented by the treatment of 'white-collar crime', and here I will compound my earlier error, if such it must be, by suggesting that the kinds of penalty imposed for violation of offshore safety regulations are derisory. To be sure, prohibition notices might, if used, provide an economically meaningful sanction, but we are still left with what will always be a number of flagrant violations, which, in my own view and apparently in that of the authorities, merit criminal proceedings. And yet, as we have seen, such proceedings as are initiated are almost invariably summary rather than solemn in nature, and even where fatalities have been involved, fines of £400 and less have been imposed. As Rusche and Kirchheimer comment, with regard to the use of fines in the general area of labour laws, such sums take no cognizance of the profits which may accrue from contravention, and the imposition of penalties becomes, in effect, a *post hoc* system of licensing.[56] Nor do I raise this point merely in order to call for the administration of 'just deserts'. I have long been struck by the extent to which the issue of deterrence, a standard item in the repertoire of law-and-order debates, so often becomes elided when it comes to the illegalities of corporations, which, *par excellence,* embody a penchant for calculation in terms of economic gain and loss. In short, I believe that penalties commensurate with the means of the companies operating in the British sector of the North Sea would have a dramatically deterrent effect upon the standard of offshore safety. But then, as one oil man explained to me, if oil companies were to be treated like criminals, they would simply leave. Maybe so. The point, however, is whether, when and if they behave as such, they should be treated differently. The argument is as old as industrialization, itself, and readers will doubtless be able to use their own knowledge of contemporary criminal peccadilloes to make the necessary substitutions in the following extract from an early nineteenth-century opinion:

> Considering the disparity between the moral culpability pehaps of getting, we will say, an ounce of tobacco or a gill of ale contrary to some excise laws, when it is visited with a ruinous penalty on some individuals not able to pay it, would it not be just and desirable to annex an adequate penalty to those who possess thousands and disregard small fines?[57]

This prompts me to one final reflection on this book as a whole. Throughout, I have been at pains, I hope successfully, to avoid claiming more for this study than the evidence and the subject can support. More specifically, I make no claim that the type of analysis which I have employed has any applicability beyond the field of North Sea oil and the arrangements which have been made for its safe exploitation. I doubt whether a study of how litter laws are enforced or (to take an example of more immediate relevance to safety) of the movement to make safety belts compulsory for children travelling in British cars would necessitate an approach of the kind which I have adopted here. All I assert is that an understanding of how the other and less creditable price of Britain's oil has been controlled requires discussion in terms of the broader issues which I have addressed. At most, I would only suggest that by happenstance, as it were, the history of offshore safety encapsulates a series of contemporary sociological issues in a particularly acute and fascinating form.

And yet, throughout the writing of this book, I have been stalked by a spectre from the past, by a broader thread of continuity which suggests there is more to this matter than the transitory exigencies surrounding North Sea oil. The reader will doubtless have noted occasional and perhaps surprising allusions to the early history of British factory legislation, a topic with which I have a passing acquaintance. These are no accident. All along, I have been struck by the extraordinary parallels which can be drawn between the history of something as up-to-date as North Sea safety and that of the earliest efforts to impose statutory control upon the operations of the 'dark satanic mills' of the nineteenth century. In that era, as in the present, there were immutable laws of capital which rendered it 'imperative' that regulation should be minimized. Then, as now, it was constantly threatened that capital would flee if subjected to any more constraints. In the nineteenth century too, it was held that those who had no practical acquaintance with the advanced technology of an industry upon which the nation's new-found wealth depended could not possibly purport to ordain how its labour practices should be ordered. It was even held that the exploitation of little children for fifteen, or twelve, or ten hours a day in textile factories was necessary, although the proponents of such a view did at least have the disingenuous decency to claim that it was also good for them. Even problems of access, of forewarning and of undue camaraderie between controllers and controlled have all rung bells redolent of issues raised more than 150 years ago. All in all, and in retrospect, a better title for this book might have been 'New Oil in Old Barrels'.

Notes and References

1. See p. 27 above.
2. Department of Energy, *Development of the Oil and Gas Resources of the United Kingdom (The Brown Book)*, London, HMSO, 1981, p. 44.
3. In 1980, the Department of Energy did not carry out a detailed survey of the offshore workforce. Instead, an estimate based on its own information and that of other Government Departments was made. The resulting figure of 22,000 included workers employed on service vessels, survey teams, construction barges etc., and, as a consequence, calculation of incidence rates for installations themselves is precluded. ibid., p. 13.
4. ibid., p. 44.
5. ibid., p. 15.
6. *Glasgow Herald,* 1 May 1981.
7. ibid.
8. *Sunday Times,* 3 May 1981.
9. Norges Offentlige Utredninger, '*Alexander L. Kielland*' – *ulykken, Nov. 1981: 11,* Oslo, Universitetsforlaget, 1981. I am deeply indebted to Ragnhild Bendiksby for her assistance in decipering the crucial parts of this report. The following description is taken from the English summary given on pp. 207–16.
10. But cf. *Offshore Safety,* (the Burgoyne Report), Cmmd. 7841, London, HMSO, 1980, p. 114.
11. *Sunday Times,* 5 April 1981.
12. Neither the Norwegian text nor the English summary makes it absolutely clear whether design and fabrication were completely satisfactory in this respect. On balance, however, the most plausible interpretation is that failure to close the doors, etc. was the issue at stake.
13. See, for example, the *Guardian,* 6 April 1981.
14. *Report on the Disaster at Whiddy Island, Bantry, Co. Cork on 8th January 1979,* Prl. 8911, Dublin, Stationery Office, 1980.
15. ibid., p. 16.
16. ibid., p. 24.
17. ibid., pp. 23–4. The statement in parenthesis has been inserted from p. 20.

18. The HSE and PED were also kept in harness 'for the time being' with regard to the preparation of unified diving regulations. This was because work on these regulations was already well advanced. Parliamentary Debates (Commons), vol. 991, 1980–1, 6 November 1980, col. 1480.
19. ibid., cols. 1479–81.
20. ibid., col. 1477.
21. ibid., col. 1542.
22. ibid., col. 1536.
23. Mr Bob Cryer, MP, for example, pointed out that many people in both the Commission and the Executive were anxious to prosecute but were 'inhibited' in this respect by the attitude of the Department of Employment. ibid., col. 1524.
24. ibid., col. 1531.
25. ibid., col. 1527.
26. ibid., col. 1513.
27. ibid., col. 1510.
28. *Guardian,* 4 June 1981.
29. The above description has been compiled from the following newspaper accounts: *Guardian,* 4, 6, 8 and 9 June 1981; *Financial Times,* 5 June 1981; *Sunday Times,* 7 June 1981. Such is the pace at which events move in connection with North Sea oil that between writing and annotating, the price was reduced by $4.25. 'The cut', reported the *Guardian,* 'represents an abrupt capitulation by the Treasury. . . .' *Guardian,* 16 June 1981.
30. See, in particular, the excellent analyses offered by John Andrews in the *Guardian,* 9 June 1981, and by David Marsh in the *Financial Times,* 5 June 1981.
31. *Guardian,* 9 June 1981.
32. SI, 1981, 399.
33. *Parliamentary Debates* (Commons), vol. 991, 1980–1, 6 November 1980, col. 1480.
34. SI, 1977, 1232.
35. *Brown Book,* 1978, p. 13.
36. *Scotsman,* 23 May 1981.
37. *The Times,* 23 May 1981.
38. *Scotsman,* 20 May 1981.
39. *Scotsman,* 14 May 1981.
40. *Scotsman,* 22 May 1981.
41. *Glasgow Herald,* 19 May 1981.
42. *Glasgow Herald,* 12 May 1981.
43. I. Taylor, P. Walton and J. Young, *The New Criminology,* London, Routledge & Kegan Paul, 1973, pp. 101ff.
44. E. P. Thompson, *Whigs and Hunters,* London, Allen Lane, 1975, p. 266.

45. W. G. Carson, 'White Collar Crime and the Enforcement of Factory Legislation', *British Journal of Criminology,* vol. 10, 1970, pp. 192–206.
46. *Parliamentary Debates* (Commons), vol. 991, 1980–1, 6 November 1980, cols. 1478 and 1536.
47. 1974, c. 37, Sections 2(4) and 2(7).
48. See, for example, *Parliamentary Debates* (Commons), vol. 991, 1980–1, 6 November 1980, col. 1536, where such complexities were held accountable for the 'protracted consultations' on this matter.
49. See above, p. 212 ff.
50. That such is the case was stated in the Commons on 6 November 1980 by the Minister of State at the Department of Energy. *Parliamentary Debates* (Commons), vol. 991, 1980–1, col. 1544.
51. *Parliamentary Debates* (Lords), vol. 315, 1971, 18 February 1971, col. 742; *Parliamentary Debates* (Commons: Standing Committee A), vol. 1, 1963–4, 19 February 1964, col. 117 and col. 98.
52. E. H. Sutherland, *White Collar Crime,* New York, Holt, Rinehart & Winston, 1949, p. 8.
53. A. V. Sheehan, *Criminal Procedure in Scotland and France,* Edinburgh, HMSO, 1975, p. 115.
54. I. Taylor, P. Walton and J. Young, 'Critical Criminology in Britain' in Taylor, Walton and Young (eds.), *Critical Criminology,* p. 30.
55. T. Mathiesen, *The Politics of Abolition,* London, Martin Robertson, 1974, pp. 14–15.
56. G. Rusche and O. Kirchheimer, *Punishment and Social Structure,* New York, Russel & Russel, 1968, pp. 174–5.
57. *Parliamentary Papers* (1831–2), XV, p. 436.

Glossary

Bale-out bottle	An emergency gas cylinder carried by divers.
Blow-out preventer	Hydraulically operated, high-pressure valves designed to maintain pressure control on a drilling well.
Buddy-line	A line connecting two divers and one which should not be capable of disconnection otherwise than by the voluntary or deliberate act of either diver.
Casing	The steel lining of a well, designed to prevent the caving in of wells, the intrusion of water etc. and to facilitate pressure control.
Cathead	A device used for driving rope by means of exerting friction pull.
Chicksan line	A flexible steel pipe with swivels which enable lines to be joined in different directions while sustaining pressure.
Collar stand	In order to drill, weight is applied above the drilling bit by means of drill collars. These collars serve the same purpose as the drill pipe in supplying the turning motion, but are very much heavier and larger in diameter. Drill collars and drill pipes are made in 30 feet lengths, but if they have to be removed from the drill hole, they are disconnected in 90 feet lengths known as stands, are moved to the side and placed vertically on the drilling floor with the upper ends fixed in racks.
Crown block	The block at the top of the derrick to which are attached the sheaves for wires used in hoisting the drilling block.
Derrick	A tower-like structure supporting the block and tackle which is used to raise and lower drill pipe, casing etc.
Derrick-man	The person responsible for racking and unracking the top ends of drill pipes etc.

Downtime	Lost drilling time due to either weather or mechanical problems.
Driller	The employee in charge of the drilling crew on the drill floor.
Drilling string	A series of jointed pipes with the drilling bit at the bottom.
Flotel	A separate floating accommodation unit.
Gin-pole	Two or more poles erected in a V-shape with the pointed section elevated and used for lifting.
Helideck	Helicopter landing area.
Hydrophone positioning control	A device allowing accurate positioning over the sea-bed by use of acoustic beacon signals.
Jack-up rig	A mobile drilling installation supported on the sea-bed by legs which can be raised to permit movement between locations.
Jacket	The primary structure of an installation constructed onshore. Deck structures etc. are added subsequently when the platform is in position.
Kelly	A shaped section at the top of the drilling string, fitting into the bushings on the rotary table and enabling it to rotate.
Monkey-board	A horizontal metal platform measuring approximately 8 feet by 3 feet and located some 80 feet above the drilling floor. This is the position from which the derrick-man carries out his part in the drilling process.
Moon-pool	A tubular hole going down through a diving vessel to the sea and permitting the lowering of diving bells.
Mousehole	A small hole in the drill floor deck for holding joints of pipe before they are added to the drilling string.
Mud-logging cabin	A cabin containing equipment for monitoring drilling mud returns and for analysing formation cuttings.
Mud tank room	The 'room' from which mud, a fluid used to cool the drilling bit, to carry away debris and to control pressure is supplied during drilling operations.
Positive displacement pump	A pump in which a piston causes a change in cylinder volume, forcing fluid out under pressure.
Power tongs	Hydraulically operated tongs for tightening or unscrewing lengths of pipe.
Purging air	Air forced through a space to dispel poisonous or dangerous gas.

Rathole	A hole drilled close to the well itself for the purpose of holding equipment such as the kelly while not in use.
Riser	The section of pipe from the blow-out preventor or well head on the sea-bed to the drilling floor level of an installation.
Roughneck	An employee who works on the drill floor.
Roustabout	A general labourer employed on such tasks as the unloading of materials from supply boats and moving them up to the drill floor as required.
Single Anchor Leg Mooring (SALM)	A means of loading oil into tankers from a mooring with one fixed leg on the sea-bed.
Saturation diving	A dive of long duration in which the body tissues absorb the inert gas in the breathing mixture and eventually become 'saturated'. The old system, known as 'bounce diving', entailed the diver going down and then being decompressed upon his return to the surface – a process which could take several hours. With saturation diving, he returns to the surface but remains in the chamber and is not decompressed. Since decompression time increases with the depth of the dive, the movement of operations into deep water made the saturation method increasingly attractive to companies because it increased diver availability. Divers may remain in saturation for a number of days.
Semi-submersible	A drilling installation with pontoons that can be flooded so that it becomes sufficiently submerged to provide adequate stability.
Shale shaker	A device for screening out formation cuttings from mud circulating in a well while drilling.
Spider elevator	A device which hoists tubulars out of a hole while gripping them by means of power operated spider slips.
Stabbing platform	A platform some 20 feet above the drill floor from which a hydraulic arm, the stabber, can guide the bottom end of the drill pipe etc. into position.
Stacked	The term used to describe a drilling rig which is laid up.
Tool pusher	The immediate superior of the driller, and the drilling contractor's chief employee on a drilling rig.

Tugger line	A steel cable attached to a winch near the monkey-board and used to assist the derrick-man in positioning the top end of drill pipe or collar stands.
Umbilical	The collection of tubes, wires etc. which supply hot water, gas, communications, electricity etc. to a diving bell.
Workover	Operations to re-enter a completed well for purposes of carrying out work to restore or increase production.

Units of Measurement:
Approximate Imperial/SI
Equivalents

Length

¼ inch	6·35mm
½ inch	12·70mm
1 inch	25·40mm
12 inches ⎫ 1 foot ⎭	304·80mm
10 feet	3·04m
50 feet	15·24m
100 feet	30·48m
1 yard	0·91m
10 yards	9·14m
50 yards	45·72m
100 yards	91·44m
1 mile	1·60km
10 miles	16·09km
100 miles	160·63km
1000 miles	1609·34km

Temperature

32 °F	0·00°C
50 °F	10·00°C
100 °F	37·77°C
110 °F	43·33°C

Weight

1 lb	0·45kg
10 lb	4·53kg
112 lbs ⎫ 1 hundredweight ⎭	50·80kg
5 hundredweight	254·01kg
10 hundredweight	508·02kg
1 ton	1·01t
10 tons	10·16t
100 tons	101·60t

Pressure

1 pound per sq. in. $6.895 \text{ kN-}m^2$

Area		Volume	
1 sq. mile	2·59km²	1 cubic foot	0·02m³
10 sq. miles	25·89km²	10 cubic feet	0·28m³
100 sq. miles	258·99km²	100 cubic feet	2·83m³

Index

Callaghan, James, 105
Cameron, T., 130, 136
capitalist system, 10–11, 298
Carrington, Lord, 163
Certifying Authorities, 277; first
    appointed, 157; challenged, 179;
    findings, 172, 177, 249; and one
    platform, *1974–7*, 242–4
'Cheviot, the Stag and the Black,
    Black Oil, The' (play by John
    McGrath) 45–6, 80
Chevron, 3
Civil Aviation Authority, 189
Claymore field, 135
Clyde field, 135
collectivism, 212–25
Committee to Review the Functioning
    of Financial Institutions, 118, 123,
    124–5
companies, *see* oil companies
Confederation of British Industry,
    110, 112
Conservative Government: *1963–70*,
    140–8, 150; *1970–4*, 100–2,
    120, 150, 161–7, 169; *1979–81*,
    110–15, 121, 129, 223
construction industry, accidents,
    21–4, 31, 41
construction, offshore: accidents,
    30, 50–2; regulations re, 156–7,
    170, 235, 242–3, 244–5
Continental Shelf: Geneva Conven-
    tion on, 140, 199, 227, 255–6,
    282; property rights in, 118,
    139–40,147
Continental Shelf Act *1964*, 118,
    126; passage through Parliament,
    139, 140–8, 150; and Scots law,
    154–5; Section *3* (criminal law)
    145–7, 192, 227, 255, 259,
    261–2, 263; Section *11(3)* 255
Continental Shelf (Jurisdiction) Order
    1968, 259, 261–2; (Amendment)
    1975, 262, 263, 283
costs of North Sea oil, 4–5, 10, 13,
    44–7, 123–5; analysis and
    safety, 77–9

crane operations: accidents
    associated with, 28, 29, 48; causes
    of accidents, 59–60, 75, 252, 253;
    jurisdiction over, 272–3, 284
crime, 'conventional', 7, 231–3,
    251, 300–2; institutional tolerance,
    7, 231–53, 300–1; *see also*
    prosecutions
criminal law, UK, 300–2; applied to
    Continental Shelf, 145–6, 227,
    255, 259; and judicial inquiry,
    261
Crown Office, 249, 263, 265–7
Cryer, Bob, 304

Daintith, T., 147–8, 227
Dam, K., 120
danger; of offshore/onshore
    employment compared, 20–30,
    38, 46; from frontier technology,
    5–6, 47–9, 50, 63, 287–8
death, *see* diving accidents; fatalities
Department of Employment, 192
Department of Energy: established,
    127, 163; administrative
    structure, 163–5; accident
    records, 17, 18–38, 172, analyses,
    49–50, 51, 71, 72, 75; *Brown
    Book*, 18, 23; *1978*, 33; *1980*, 3;
    1981, 39, 41; employment figures,
    18–19, 22, 40, 41, 303; Develop-
    ment and Production Programme,
    127; and North Sea oil develop-
    ment, 106–7, 127–8; relations
    with oil industry, 16, 169,
    174–81, 187, 201, 204–11,
    291–2; responsibility for North
    Sea safety, 7, 155–6, 163–7, 169,
    187, 192, 290–2, 293–4, 295,
    298, 300; enforcement of safety
    regulations, 235–7, 239–53; and
    other Government Departments,
    187, 192; and Health and Safety
    Commission, 187–8, 190–7,
    202–11, 222–5, 290–2, 294; and
    Scots law, 264–7, 270; and

union organization, 220–1,
222–4
Department of Energy, Petroleum
Engineering Division (PED):
established, 163; structure, 164–5;
staff, 205, 208–9; *1980*, changes,
290–1; Inspectorate, 172, 174–5,
194–5, 206, 236–7, 239–47,
291; safety expertise, 173–4; and
Health and Safety Executive,
202–5, 290–1; and safety regulations, 173–6, 178–81; staff
205, 208–9
Department of the Environment,
189, 192
Department of Health and Social
Security, injury statistics, 38, 40
Department of Trade and Industry:
1973, responsibilities and
structure, 162; accident records,
18–19, 33; and development of
North Sea oil, 91–2, 100, 122,
125, 126–7, 162, 165; and safety
in North Sea, 151, 172, 189, 194,
197; Marine Division, 18–19, 189
Derwent, Lord, 141, 146
Det Norske Veritas, 277
divers, employment of, 19, 270,
278–80; regulations re, 284, 285,
294
diving: and pressure, 76; regulations,
157, 284, 285, 294–5; saturation,
308
diving accidents, 28–9, 50, 270;
fatalities, 23–5, 31, 42–3, 47,
63–70, 82; jurisdiction over,
259–60; prosecutions over,
267–70, 272, 275–80, 285;
dynamically positioned vessels,
24, 78
domestic workers offshore, 28, 29
Doran, Charles, 46
drilling, accidents involving, 28, 29,
47–8; causes of, 52–6, 74
Dyce Airport, Aberdeen, 4, 45

economics, 10, 46, 77–8; of
Britain, 2–3, 10–11, (*1960s*)
85–8, 90–5, (*1970s*) 95, 98–102,
103–16, 129; determinism, 12;
international finance, 116–18,
123–5, 200, World, 10–11, 85,
88–90, 95–8, 102–3, 110,
116–17; *see also* costs of oil;
financing North Sea development;
taxation
Edwardes, Sir Michael, 112
Ekofisk Bravo, 95; blow-out, 5, 42,
71, 195
Emery, Peter, 155
employment: numbers employed
offshore, 18–19, 22, 303; relationships defining, 270, 278–80,
284; of divers, 19, 270, 278–80,
285, 294
Employment Protection Act *1975*,
213, 220–1
energy crisis, world, 95, 102
Errol, F. J., 141–2
Esso, 178, 201
European Economic Community,
88, 113, 198–200; Advisory
Committee on Health and Safety
at Work, 198
expertise: legal, 248; safety, 172–5,
298; technical, 128, 169–71
Exxon, 90

Factories Acts, 67, 191; devised,
155; passing compared to North
Sea legislation, 302–3; proposed
to apply to Continental Shelf,
145–6, 148
Factory Inspectorate, 205–7
Fastnet yacht race, 42, 80
Fatal Accidents Inquiry (Scotland)
Act *1895*, 259, 261
Fatal Accidents and Sudden Deaths
Inquiry (Scotland) Act *1906*,
259–60

Operational Safety, Health and Welfare, 157–8, 202
offshore/onshore employment compared, 38; accidents, 25–7; fatalities, 20–5, 46; causes of these, 48–9
Organization of Petroleum Exporting Countries (OPEC), 10, 91; formation, 90; and British companies, 100; and oil prices, 2, 95, 96, 97–8, 101, 102
oil: in world economy, 85, 88–91; prices, 2, 89–90, 95–8, 101, 102, 111, 293
oil, North Sea: discovered, 2, 95, 99–100, 101; supplies, 3, 108; development policy, 85, 90–5, 100–2, 105–16; speed of development, 9, 10, 11, 84–5, 92–3, 106–8, 112–13, 135–6, 169, 242, 291–2, 297; and relationship of British Government/oil industry, 116–31; financing, 4, 115–25, 129–30; revenue from, 3, 108, 114; effect on British economy, 88, 102, 105–6, 108–9, 111–12, 115; *see also* taxation
oil companies, multinational: relationship with British Government, 116–31; effect of this on safety regulations, 169–81; finance, 118–19, 122–5, 128–30; ruthless image of, 43, 45–7, 48, 72, 76–8
Ormerod, Paul, 104–5

Parliament: debates Burgoyne Report, 290–2; Committee to Review the Functioning of Financial Institutions, 118, 123, 124–5; and Continental Shelf Act, 139, 140–8, 150; Mineral Workings (Offshore installations) Act, 151–3, 190; Public Accounts Committee, 100, 120, 122, 126, 136; *see also names of*

Acts of Parliament; Conservative Government; Labour Government
Petras, James, 14, *225*
Petroleum (Production) Act *1934*, 144, 147, 149
Petroleum and Submarines Pipelines Act *1975*, 32, 33, 130, 230
Phillips, *95*
pipelaying: accidents, 31, 33, 260–1; regulations, 32, 157, 235; inspection, 246
pipelaying barges, jurisdiction over, 32–3, 237–8, 256–7, 258, 259–61
pipelines, 4
police: and Continental Shelf Act, 255; Scottish, function of, 264, 266, 300; jurisdictional powers, 255, 256–7, 261, 282
pollution, 5
Prescott, John, 220–1
press, the, 42–3
pressure, on oil workers, 71–9
private enterprise, 120–1
Procurators Fiscal, 51, 248; functions of, 81, 264; and fatalities inquiries, 258–66, 269–71, 274, 277–9; and prosecutions, 249–51, 300
production phase of oil industry, 19–20, 28, 29–30
property rights in Continental Shelf, 118, 139–40, 147
prosecutions, 247–51, 264–7, 281; examples of, 267–80; decision to prosecute, 264; and fatal accident inquiries, 265–7
Public Accounts Committee, House of Commons, 100, 120, 122, 126, 136
Purdy, David, 11, 85, 87, 133

quarrying: accidents, 21, 22, 23, 24, 25, 26, 39; Inspectorate, 25, 40, 192, 207–8